GOOD NEWS FROM THE FRON

SOLVIVA

HOW TO GROW $500,000 ON ONE ACRE & PEACE ON EARTH

LEARNING THE ART OF LIVING, WITH SOLAR-DYNAMIC, BIO-BENIGN DESIGN

REVEALING THE TRUTH
about how we can provide electricity, heating, cooling,
transportation, food, solid waste and wastewater management
in ways that reduce pollution and depletion of resources
by 80 percent or more, and that at the same time
reduce cost of living and improve quality of life.

by

Anna Edey

TRAILBLAZER PRESS
Martha's Vineyard • Massachusetts

All photographs and illustrations are by Anna Edey unless otherwise noted.

The author is not responsible for the level of success that others may achieve by adopting the designs and methods described in this book. Results are entirely dependent on levels of judgement, accuracy, intuition and effort applied.

Library of Congress Catalog Card Number 97-91422
ISBN 0-9662349-0-1

First Edition

All Rights Reserved
Copyright © 1998 by Anna Edey

Published by Trailblazer Press
RFD 1 Box 582, Vineyard Haven, MA 02568
Tel./Fax.: 508-693-3341

Printed on chlorine-free paper with high recycled content.

Thank you

I want to express my deepest gratitude to all the many who helped create the Solviva work and this book.

To Lisa, Kirsten and Annika, my daughters, for putting up with my relentless commitment to this work, and for their multifaceted help and intelligent critiques along the way.

To Mait Edey, my ex-husband, for his continuing encouragement.

To Helen Edey, for providing information (such as the BTU-power of rabbits) without which the Solviva greenhouse would probably never have been created.

To Maitland Edey, who, even after passing way, has remained my mentor.

To my mother and father, for a childhood filled with opportunities for discovering the mysteries and wonder of life on Earth.

To Bo Jufors, my brother, for being my soul mate and fellow seeker/finder, and for all his help.

To Margaret Lloyd, David Rosenmiller, Donna Goldberg, David Wright and many others for providing financial assistance without which the Solviva greenhouse would not have reached fruition.

To Gary Krogseng, for providing the 3M Sungain glazing and ongoing assistance.

To Robert "Sardo" Sardinsky, without whose encouragement, energy and skills the greenhouse would never have been created.

To Jonathan Schaull, Neil Withers, John Strunk, Willy Reid, Barry Downes, Donna Byers, Donna Grimes, Bruce Fulford, Earl Barnhart, Jenny Ruffing, Laurie Boosahda, Rafe Brown, Ruthie Dreier, Barth Jarek, Kevin Williams, Jeff Kinzel, K-K and Dan Cohen, and all the many other angels who helped to create the Solviva greenhouse.

To Ed Klugman for understanding the vision and enabling many to learn from hands-on experience at Solviva.

To Deborah Huntley, chef of the Harbor View Hotel, and all the other chefs of excellent restaurants on Martha's Vineyard and around Boston, for being loyal and appreciative customers of Solviva Salad.

To Peggy Vance, Featherstone Meetinghouse Center for the Arts, and Farm Neck Foundation for enabling the installation of the first Solviva Biocarbon Filter septic system out in the "real world".

And, most important for the final gestation and birth of this book, to Rachel Orr for extraordinary organizational editing, and to Marianne Goldberg for intelligent line editing and Pia Webster for meticulous copy editing, and to all the many friends who read the developing manuscript over the years and offered insightful comments. And thanks to so many others who have helped in so many different ways over the years.

We are all one, interconnected, interdependent, just as all the atoms on Earth are interconnected and interdependent. Together we can discard those habits and technologies of the 20th century that are clearly harmful, and embrace those that have proven to be beneficial and sustainable, to create a Golden Age in the 21st century and beyond. Together, we Homo sapiens can learn the art of living with Life.

Table of Contents

Introduction 9

Some Current Realities 13

Part 1: The Truth Revealed in Color 17

Part 2: A Visit to Solviva 57

Winter in My Home. Winter in the Solviva Greenhouse.
Summer in the Solviva Greenhouse. Summer in My Home.
Summary of the Advantages of Living the Solviva Way.

Part 3: How I Got On the Path of Seeking Better Ways to Live and What I Discovered Along the Way 66

Three Catalysts for Change
No.1: Urine Power
No.2: Meditation
No.3: Fire

Heating 78

Cooling 83

Wastewater Management 85
The Solviva Compostoilet. Graywater Purification.
The Solviva Composting Flush Toilet.
Easy Composting. Solar Pasteurization.

Food Production 95
The Greenhouse Dream. Actualizing the Dream. Animal Power.
The First Winter. The Greenhouse Cornucopia. Trouble in Paradise.
The Earthlung Biocarbon Air-Purification Filter. Birth of Solviva Salad.

Electricity 107
Solar Electricity

Transportation 112
My Electric Car

Solid Waste Management 115
Comprehensive Recycling

Part 4: A Call to Action: Proposals for Jump-Starting a Better Future 119

1. The Greening of the White House.
2. A School that Protects our Economy, Health, Environment, and Resources, Today and Tomorrow.
3. An Ecommunity Recreation Center.
4. A Solviva Restaurant and Business Center.
5. Solid Waste Management that Results in 90 percent Recycling.
6. Public Transportation.

A Tale of Two Cities
Grayberg or Greendale: Where Would You Rather Live? 137

Part 5: A Collection of Powerful Quotes 143

Part 6: Construction and Maintenance of the Solviva greenhouse and Farm
What I Did and What I Reccommend. 155

The Solviva Designs: Review, Evaluation, Recommendations 166
The Foundation. The Wood Framing. The Ventilation. The Paths.
The Growing Beds. The Hanging Growtubes. The Watering Systems.
The Solar Heat Storage Systems. The Backup Heating System.
The Earthlung Filter. The Attached Cold Frame. The Glazing System.
The Pitch of the South Roof. The Walk-in Growshed.
Cost of Materials. Building a Greenhouse in a very Cold, Dark Place, such as Alaska.

Solviva Greenhouse Management 173
Insect Management.
13 Golden Guidelines for Minimizing Pest Problems.

More Detailed Description and Recommendations of the Solviva Management Techniques 186
Composting. Soil Fertility. Outside Garden Preparation. Mulching.
Seeding. Transplanting. Planting. Weeding. Watering. Harvesting.
Washing, Draining, Packing. Marketing.

A Year at Solviva 201

The Animals 205
The Chickens
The Rabbits
The Sheep

Part 7: Recommended Resources 219
Favorite Seed Varieties for Solviva Salad
Suppliers of Seeds and Goods for Sustainable Living
Recommended Reading
Conversion Tables

I Pray

❦ ❦ ❦ ❦

I Pray
that we will come some day
to realize that "they"
are not the ones to blame.

We'll see
it's up to you and me
to make democracy
our responsibility.

Though we say the western world is
civilized and great,
now we know that living thus
is threatening our fate.

Poisons course through every vein
of every living being,
through air and earth and water too,
there's almost nothing left that's clean.

Wake up,
do right,
speak up for what you feel is right,
use your mighty heart and mind.

Live truth,
do good,
then the way you live is right,
and you will be a guiding light.

We can all have what we want
and what we need in life
without destroying life on Earth
and plundering all its worth.

We can live on daily income
paid out by our sun,
instead of robbing resources
from our daughters and our sons.

We can
live right
we do have everything to gain,
and all to lose if we don't change.

I pray
that we will come some day
to realize that "they"
are not the ones to blame.

We'll see
it's up to you and me
and everyone we see
to learn the harmony.

We'll see
it's up to you and me
to live the way we want
in peace and harmony.

Anna Edey, 1990

1960's

1970's

1980's

1990's

Kirsten Edey

I dedicate this book to my grandchildren
and to all other children,
for they shall inherit the Earth.

When they grow up,
life on Earth will be either far better or
far worse than it is today.
The choice is up to us, the adults.

INTRODUCTION

❦ ❦ ❦

"In Our Every Deliberation,
We Must Consider the Impact of Our Decisions
on the Next Seven Generations."

From the Great Law of the Iroquois Nation

Over the millennia, we Homo sapiens, the so-called Wise Ones, have been striving to improve the quality of our lives and to ensure an ever better life for our children. Many cultures around the world, including Native Americans, consciously planned and acted to protect the well-being of seven generations into the future.

In my native land, Sweden, I know a carpenter who uses wood that was milled and stored by his father from trees that were girdled at the right time of year and cut by his grandfather. The trees were trimmed, thinned and protected by his great-grandfather, great-great-, great-great-great-, and great-great-great-great-grandfathers. The seeds for those trees were collected from the best quality trees and sown on land that was cleared, cultivated, fertilized, weeded and watered by his great-great-great-great-great-grandfather, seven generations back, and he in turn was using wood from trees that were produced during the seven generations before him. Will today's carpenter sow and tend the seeds for his future seventh generation? Not likely.

Only during the last century, especially since World War II, have we all but forgotten the concept of planning for the seventh generation. Forests are being razed; soil, groundwater and lakes are being polluted and depleted; oil is being burned - all at an unprecedented and utterly unsustainable rate, as if there will be no tomorrow. Nuclear wastes, so toxic that a few pounds can kill a city, are accumulating by the thousands of tons, with no viable solution for their disposal. Most of our food is sprayed with many different toxic chemicals, many of which become far more toxic as they recombine in our bodies and our environment.

The cost of living has gone up astronomically, but we hardly feel it because most of the costs are hidden in waste, depletion and debt deferred to the future, to burden our children and future generations. The U.S. government has accrued a debt that amounts to some $20,000 for every man, woman and child in this nation. This is on top of the enormous personal debt owed by most people for credit cards, bank loans or mortgages. According to the War Resisters' League, 50 percent of our taxes are spent to pay for past and present military budgets.

Perhaps the pivotal reason for the lack of appropriate care for the future is The Bomb. Ever since August 6, 1945, it has become self-evident that our technology has evolved to the point that we can destroy our own species and most of the rest of life on this wondrous planet. I am baffled by the sense of relief that spread over the U.S. when the USSR fell apart: "No more cold war, no more danger of the bombs!" Yet, thousands of nuclear bombs still sit in rusting silos, now controlled by smaller unstable nations, capable of going off at any moment, triggered by failing technology or a vengeful terrorist.

It seems that we are programmed to self-destruct with either a bang or a prolonged agonizing whimper in the near future. No wonder stress, waste, crime, violence, disease and escapist behavior are rampant. No wonder people no longer think seven generations ahead.

So what are we to do?

Shall we believe those who claim the situation is hopeless, that we have gone to the point of no return and therefore we might as well keep doing more of the same?

Or shall we believe those who say we may be able to save ourselves, but only if we drastically change our lifestyle and standard of living? According to these people, we must give up many of the comforts and conveniences that we hold dear, such as cars, long hot showers and deep baths, meat, plastics. Some claim that giving up toxic pesticides will lead to ugly, wilted and scarce food in our markets and to worldwide crop failures and starvation. Saving the environment will supposedly lead to the loss of millions of jobs. Spotted owls or jobs, snail darters or electricity. If such tradeoffs are required in order for us to survive, then our prognosis is gloomy indeed.

But in truth no such tradeoffs are required. This is a book of Great Good News. I will demonstrate that we now, today, have the technology and know-how to reduce pollution and depletion of resources by 80 percent or more, and I will show how this can be done in ways that can reduce our cost of living and improve the quality of our life.

Since 1977 I have accumulated evidence through direct experience on my farm on the island of Martha's Vineyard, off the coast of Massachusetts. Here I have been seeking to find ever more harmonious ways to live on Earth. The more I looked at the problems caused by our modern ways of living, the more I realized how profoundly these problems are all interconnected. My focus has therefore been broad and comprehensive, to design more sustainable and economical ways to provide for the basic necessities of our lives, including heating, cooling, electricity, food, transportation, and management of wastewater and solid wastes.

It was in the late 70s that I began to develop my own versions of what I call solar-dynamic, bio-benign living design, and I have been continuing this work ever since. These ongoing experiments have provided results that proved far more successful than I had ever imagined possible. For instance, who could have predicted that sewage can be filtered through leaves and wood chips and in five minutes be transformed into odor-free water containing 90 percent less nitrogen? I would never have thought it possible - until I did it. Who could have predicted that tomato plants could grow 30 feet long and live four years right in the kitchen, without any pesticides or normal fertilizers, producing superb-quality tomatoes continuously, even in the middle of winter? I certainly did not think this would be possible - until I did it.

Based on my experiences and knowledge I conclude that with today's technologies the following is possible:

1. We can manage our wastewater, from homes, schools, business and industry, in ways that eliminate water pollution, thereby protecting our drinking water, fishing industry, wildlife, ponds and harbors - and this can be done in ways that save money, as well as irrigate and fertilize our landscapes and forests.

2. We can recycle 90 percent of our solid wastes in ways that save time and money, energy and resources, and that greatly reduce pollution - while creating more jobs.

3. We can produce high yields of high-quality organic foods year-round in any climate, in urban and rural locations, without heating fuels or cooling fans, without toxic chemicals, and with far less irrigation water. Thus we can greatly reduce the depletion and pollution of soil, water, oil and other resources, as well as avoid the health hazards caused by conventional food production methods.

4. We can use solar power to provide most of the energy to heat and cool our homes, schools and other buildings, with renewable plant-derived fuels as backup. This can create more jobs and reduce by 80 percent or more the cost, pollution and depletion caused by conventional methods that rely on oil, gas, and coal.

5. We can greatly reduce our consumption of electricity with various efficiency technologies, and most of the remaining requirements can be satisfied with small-scale solar, wind and water power sources, thereby reducing by over 80 percent the current use of oil, coal, gas, and large-scale hydro - and eliminating nuclear power.

6. We can greatly improve public transportation, and both private and public transportation can be provided with electric vehicles with batteries powered primarily by the sun, supplemented with methanol and other plant-derived fuels. This can reduce consumption of gasoline by 80 percent or more.

There are many in the U.S. and around the world who share my views that these goals are possible to achieve, but most find it unbelievable. I can understand this skepticism. After all, very few people have had a chance to experience such sustainable technologies firsthand. Instead they hear that solar was proven in the 70s to be ineffective, that recycling can never exceed 30 percent, that electric cars are slow and have a very short range, and that eliminating pesticides would result in unpalatable food and global famines.

I am not saying that the transition to sustainable methods will be easy. There are formidable forces at work to maintain the current infrastructure, among them: inertia, doubt, lack of knowledge and experience, pessimism, fear, greed, and perhaps most obstructive, vast investments in the status quo and paralyzing regulations and bureaucracy.

But I believe most people are good at heart and want to do their best to help ensure the well-being of all, for now and for the future. Unfortunately, we are not yet at the point where all the better alternatives are easily available. There is a Catch-22: the sustainable technologies will not be mainstreamed into society until masses of people want them, and masses of people will not want them until they are mainstreamed. There will be a phasing-in period before these technologies become widely and easily available, and are thus able to reach their astonishing cost-effective potential.

As more and more people learn about better ways to live and have a chance to experience firsthand that these methods decrease costs, pollution and depletion and are beautiful, reliable and convenient, then a critical mass of this knowledge and understanding will be reached. When this happens, hope for a better future will be rekindled, and with it the will to spread sustainable and harmonious living systems - like a blessing around our planet.

With this book I hope to greatly increase general public awareness of what is possible for us to accomplish with today's technology. I wish you an inspiring journey through the words and pictures herein, and happy envisioning, planning and action for a better future for us all.

Anna Edey

Thanksgiving, 1997

SOME CURRENT REALITIES

It is a thrill to fly back home to Martha's Vineyard, especially when lucky enough to sit in the copilot seat in the tiny commuter plane. After passing Boston, the smog clears and the view becomes captivating as we bump and drone south along the coast of Massachusetts.

The landscape looks pristine from above, neat ribbons of highways with cars traveling like ants. Even the dumps look clean and tidy, and the wisps of smoke from the power plants and incinerators appear so innocent. Only the ponds, lagoons and inlets offer any visible indications that all is not well. Most of them look like festering sores, surfaces partially covered by brown and green algae infestations. These unhealthy conditions are caused by the influx of excess nutrients that for years was blamed on water birds and on runoff from agriculture and lawn fertilizers. However, it is now known that algae pollution is primarily caused by the nitrogen that seeps with the groundwater from conventional septic and sewage systems, whether nearby or distant from the body of water. In other words, the algae pollution is caused mostly by human body wastes.

Some 4 million people live in eastern Massachusetts, and their bodies release wastes that contain about 40 million pounds of nitrogen and 15 million pounds of phosphorus per year. Standard septic systems are incapable of reducing this nitrogen to any significant degree, and thus release some 35 to 55 ppm (parts per million) of nitrogen into the groundwater. Many central sewage treatment facilities release about 25 ppm. Thus 20 to 25 million pounds of nitrogen a year flow with the groundwater, unabated, at a rate of about 1 to 3 feet per day, to the nearest surface waters, whether they are 100 feet or 10 miles away.

Levels of nitrogen above 10 ppm in drinking water are dangerous to human health because they can reduce the blood's oxygen exchange capacity. (Methemoglobanemia is one of the causes of the dreaded sudden infant death syndrome, or SIDS.) However, levels of nitrogen even lower than 10 ppm are harmful to the environment because they cause massive algae growth, like a fungus infection, in ponds and lagoons, rivers and harbors. This leads to foul odors, stagnation and rapid eutrophication, and to pollution and death of shellfish beds and spawning grounds.

On this cold day in late November, these same 4 million people are burning roughly 15 million gallons of oil in power plants, furnaces and heaters in order to have warmth and electricity for their homes and places of learning, work and recreation. This fuel was extracted in faraway places such as the Persian Gulf region, Northern Atlantic, Nigeria, Alaska, Mexico and Venezuela, with disastrous ecological consequences, and transported across the oceans in giant oil tankers, resulting in annual spills that total many times more than the Exxon Valdez spill.

Burning one gallon of oil releases 5.5 pounds of carbon, even with the best catalytic converters or filters, and this carbon combines with oxygen in the air to form almost 20 pounds of carbon dioxide (CO_2). Thus the 15 million gallons of oil that just these 4 million human beings are burning on just this one day in November is resulting in roughly 300 million pounds of CO_2 which rises into our atmosphere.

In addition, on this day roughly 1 million cars, buses and trucks are consuming about 5 million gallons of gasoline, causing 100 million pounds of CO_2 pollution. Also, the vehicles emit dozens of other harmful substances which form the thick gray-brown haze that hangs heavy over greater Boston.

Worldwide, over 5 billion tons of carbon are released annually into our atmosphere through chimneys, smokestacks and tailpipes. This carbon combines with oxygen to form some 20 billion tons of CO_2 to add to our atmosphere. All this is above and beyond the CO_2 that is naturally emitted by life forms and volcanoes. Scientists still disagree on precisely what will result from all this man-made CO_2, which has already increased atmospheric CO_2 levels more than 25 percent above preindustrial levels. Predictions range from a new ice age to global warming and a 30-foot rise in sea levels. But there is general agreement that it will have, and indeed is already having, serious effects, as evidenced by the increasing frequency and violence of devastating hurricanes, floods, droughts, and fires.

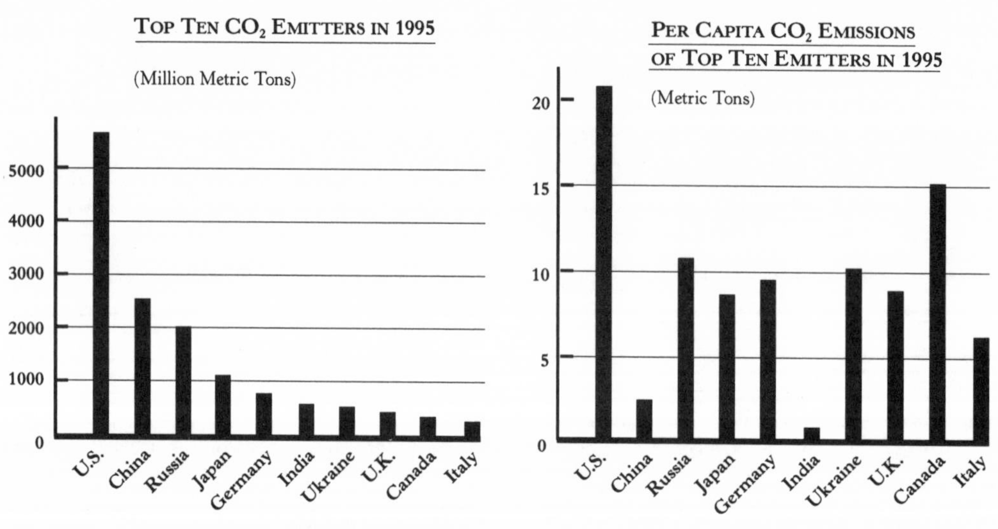

Source: Oak Ridge National Laboratory, 1997

These same 4 million people in Eastern Massachusetts are almost completely dependent on food that is trucked and flown in from far away. The average distance food is shipped from place of production to place of consumption is 1,200 miles. Most of these foods have been grown in ways that deplete vast quantities of oil, water and soil. In addition, the prevailing agricultural practices threaten the health of agricultural workers and consumers as well as of our environment. About 2 billion pounds of some 300 different varieties of pesticides are manufactured in the U.S. annually, about half of which are applied in the U.S., while the rest are exported. Some of the exported pesticides are now illegal in this country because of their high toxicity, but they are used to grow food in other countries and then, in spite of spot checking, return to us in the foods we import.

Thus, much of the food that people consume contains remnants of harmful herbicides, fungicides, insecticides, and other life-damaging substances, many of which become far more toxic when combined with others. Furthermore, most of this food has been grown with only a few man-made chemical nutrients, in "dead" soils lacking the full spectrum of life-promoting natural substances that can be had only with compost. We may well wonder how much these agricultural practices are contributing to increasing incidents of various health problems, such as allergies, asthma, cancer, hyperactivity and immune deficiencies.

These same 4 million people generate roughly 3 million tons of household solid wastes per year, as well as many more millions of tons of industrial wastes, consisting of various metals, plastics, papers, glass and food. At least 90 percent of these wastes are actually precious resources that are recyclable, but instead, most are buried in landfills or burned in incinerators, causing tragic expense, pollution and depletion.

The plane begins to descend as we approach the southern coast of Cape Cod, and suddenly my home community, the island of Martha's Vineyard, appears on the horizon. It lies in the Atlantic Ocean about seven miles off the coast of Massachusetts, consisting of 100 square miles of rolling hills of terminal moraine and flat stretches of outwash plain left by the last ice age, about 10,000 years ago.

This island is located just north of the 41st parallel, which means that on the shortest day of the year the sun rises no higher than 25 degrees above the horizon, and there are only nine hours between sunrise and sunset. Furthermore, winter is not only the time with the shortest days and the lowest light intensity, but it is also the cloudiest period. During the course of the year, the Vineyard receives 49 percent of possible sunshine (compared with over 90 percent for Arizona), but we sometimes have five to six weeks in a row between November and January with less than 20 percent of possible sunshine, maybe two half-days of sunshine per week.

Winter temperatures can dip below zero degrees F with a far lower windchill factor because of frequent high winds. Summers are hot and humid, sometimes exceeding 90 degrees, and at such times the winds are a blessing. Several times a year gales exceed 50 mph, and hurricane force winds registering 80 to 90 mph have occurred at least six times within the last 15 years.

Annual rainfall is about 45 inches, but droughts of several months are not uncommon in summer. The soil is sandy and relatively low in organic matter. It is naturally acid and increasingly more so due to acid rain.

The Vineyard has a "sole source aquifer", meaning we have only one source of drinking water. This aquifer is replenished only by rain and snow, which percolate down through layers of soil and sand to join the lens of fresh water that permeates gravel layers below. And yet, in spite of the fact that this is indeed the only source of drinking water, the year-round and seasonal residents on the Vineyard flush down their toilets some 300,000 pounds of nitrogen annually (equivalent to the amount of nitrogen contained in 150,000 standard 40-pound bags of 5-10-5 fertilizer) and 100,000 pounds of phosphorus. This, together with various household cleaning chemicals that go down the drain, enters thousands of septic systems and from there leaches into the aquifer, along with toxic chemicals from various other sources such as dumps and oil tanks.

As we swing in over the island, I catch a glimpse of my home, the Solviva Organic Farm. The gardens and fields lie dormant in muted winter colors, and the sun glints off the shining 100-foot-long crystal that lies in their midst. This is the Solviva Winter Garden solar greenhouse, and inside is a thriving garden capable of producing some 1,600 organic salad servings per week. Amazingly, even in the coldest conditions, this greenhouse stays warm enough to remain highly productive without any heating fuel, while other "normal" greenhouses require thousands of gallons of oil or gas, or many cords of firewood.

By the time I arrive home, it is dark and very cold outside, but inside it is still toasty warm from the day's sun. As I turn on the lights, I see that the plants in the indoor garden look happy, as always. During the two cold weeks that I have been away, this house has stayed warm enough to keep the plants healthy without any heating other than the sparse winter sun. I pick one of many sweet juicy tomatoes that ripened while I was away.

It gives me an extraordinary sense of freedom and satisfaction to be able to go away for an extended period of time without the preparations that must be done in a normal home, such as draining the plumbing and boarding the plants with friends, or leaving the heat on (with the inherent risks) and asking a friend to water. To some people, the fact that I can safely leave home without elaborate preparations is one of the most remarkable advantages of my home.

PART 1

THE TRUTH REVEALED IN COLOR

I had a dream of learning to live in harmony with life on Earth, in ways that cause far less harm to our water, soil and air, to our health, economy and security. I wanted to live in ways that help protect the environment and resources for many future generations.

In 1980 I had the opportunity to start actualizing this dream, and before long it became evident that my designs and methods were succeeding far beyond my highest hopes.

This book is about what I discovered and what the implications are for individuals, communities and nations around the world.

THE SOLVIVA SOLARGREEN HOME

My home was completed in 1981. Here it is a few years later in a raging blizzard, at 4 degrees below zero F. Inside it is comfortable with only wastepaper burning in a Franklin stove.

Next day it is sunny but just as cold, and inside it is cozy with only solar heat. The tomato plants are four years old, with 30-foot vines winding along the long skylight. They produced superb quality tomatoes continuously, even through the long, dark, cold New England winters. The garden room is fully open and integrated with the kitchen, living room and weaving studio, and even the upstairs.

The abundant salad garden in mid-winter. Note the window insulation panels in the storage place below the windows. At night it takes a minute or two to put the panels on the windows to prevent the warmth from escaping.

Ruthie at the loom.

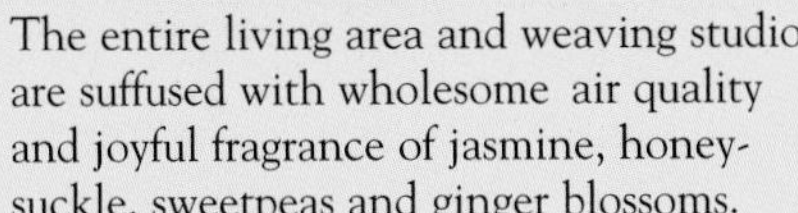

The entire living area and weaving studio are suffused with wholesome air quality and joyful fragrance of jasmine, honeysuckle, sweetpeas and ginger blossoms.

These tomatoes, so succulent and sweet even in mid-winter, brought tears of joy and hope to everyone who came to visit.

THE SOLVIVA GREENHOUSE and FARM

The success of my solar home designs inspired me to dream of a larger greenhouse where more people could come to experience the proof that we can live sustainably. I dreamed of resident chickens and rabbits to supplement the solar heat, hoping to eliminate the need for any heating fuel. I hoped for self-cooling without any fans, and pest control without any toxic pesticides.

In 1983 the Solviva Winter Garden greenhouse started rising.

Robert Sardinski

Robert Sardinski

Here we are tilling the best of goodies into the soil.

Robert Sardinski

Jonathan artfully building the raised beds.

Robert Sardinski

Gary, Jonathan and Kevin applying four layers of 3M Sungain glazing.

Sardo completing the solar electric installation to power small fans and pumps to circulate solar heated air and water.

Ruthie Dreier

The waterwalls absorb enormous amounts of solar heat during the day and release it at night, without substantially reducing valuable growing space.

The completed Solviva Winter Garden greenhouse.
Outside: 4 degrees below zero F. Inside: 80 degrees F.
Even under the coldest conditions this greenhouse proved capable of staying warm enough without any heating fuel.

A relaxing lunch break for Jonathan, Jenny, Gary, Donna, Kevin and Barry, a few of the many angels who helped actualize the Solviva greenhouse dream.

Katherine Rose

The tub stores 600 gallons of solar-heated water. It also makes a great hot tub. Note the chimney connected to a firechamber under the tub - just in case back-up heat would ever be required.

Ruthie Dreier

Ruthie Dreier

Nasturtium and kale. Mache. Ruby Swiss chard. Tah Tsai.

THIS GREENHOUSE PROCEEDED TO FAR EXCEED ANY PREVIOUS RECORDS OF ENERGY EFFICIENCY, PRODUCTIVITY AND NON-TOXIC MANAGEMENT, FAR BEYOND EVEN MY OWN HIGH HOPES.

A cascade of luscious melons on 30-foot vines.

This collard plant has leaves so big that one is enough to feed a family. After 3 years it had grown to 5 feet and finally went to seed, and was then cut down with a chainsaw.

Perhaps most amazing of all: this is one chard plant that grew for 3 years before bolting, producing about 100 leaves per week on 30 plantlets that developed on side branches that spread 5 feet high and wide.

Thousands of nasturtium blossoms on 30-foot vines.

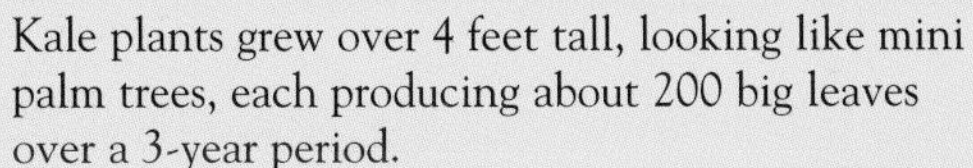

Kale plants grew over 4 feet tall, looking like mini palm trees, each producing about 200 big leaves over a 3-year period.

Donna under one of several bowers of 25-foot tomato vines dripping with many hundreds of tomatoes.

I had always assumed that tomatoes would be the main crop in the Solviva greenhouse, but they were soon eclipsed by another product, far more predictable and profitable.

Solviva Salad

I began to market a product never before offered commercially in America: a ready-to-serve blend of perfect young leaves of many varieties of salad greens.

Snow lies on the lower glazing and it is below zero degrees F outside. Inside it is 80 degrees, without any heating fuel. Here are a few of the 50 different varieties of salad plants that grow with exceptional rate of productivity.

Soon orders for Solviva Salad were coming in from restaurants and markets faster than we could produce. How could production be increased without building another greenhouse?

The solution was to make hanging gardens, using the vertical air space within the greenhouse, by hanging seven levels of growtubes.

This shows the total of **nine growing levels** within the 18-foot high Solviva greenhouse: the raised beds, the seedling nursery by the upper catwalk, and seven levels of growtubes.

Highest quality lettuces growing in 4-inch diameter growtubes.

HARVESTING THE ABUNDANCE GROWING IN THE GROWTUBES AND THE RAISED BEDS IN MID-WINTER.

Bruce harvesting the downstairs growtubes.

Rafe Brown

Marianne harvesting the upstairs growtubes.

Tara washing and draining the greens.

Rafe Brown

Annika showering the greens to keep them crisp during harvest.

Portrait of a 1-ounce Solviva Salad.

Continuous waves of seedlings being produced in the Solviva Growshed in spring, summer and fall.

Continuous seedling production at the catwalk level in the Solviva greenhouse.

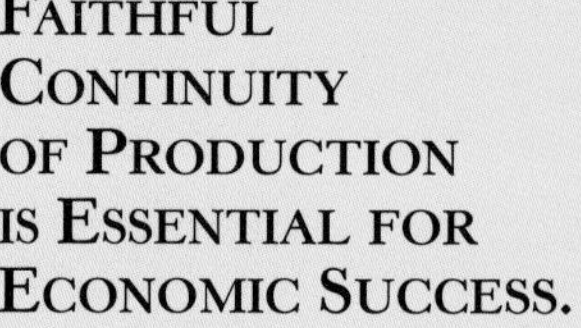

FAITHFUL CONTINUITY OF PRODUCTION IS ESSENTIAL FOR ECONOMIC SUCCESS.

The outside garden in winter.

The outside garden in summer.

The Solviva greenhouse in a blizzard, containing a thriving garden capable of producing 1600 salad servings per week, without any heating fuel or harmful pesticides.

The Solviva greenhouse is also exceptionally productive all through the summer, without any cooling fans. Here it is reflected in the pond, along with the **photovoltaic panels** which produce the electricity to power the small fans and pumps that circulate the solarheated air and water.

Insect pests are controlled by 13 Golden Guidelines, including beneficial insects. No toxic pesticides are used, not even those that are considered safe by normal organic standards, such as rotenone, pyrethrin or insecticidal soap. After all, they are meant to kill, and will certainly do harm also to the beneficial insects and to the whole ecosystem.

Flowering plants are essential for providing habitat, pollen and nectar for beneficial insects. Here are dill, pineapple sage, marjoram, rosemary, Johnny-jump-up, cilantro, fennel, borage, miscellaneous crucifera.

The soil is kept fertile and alive with additions of compost made from the bedding of the resident chickens, rabbits and sheep. This ensures conditions that promote an abundance of different life forms that contribute to controlling pest insects and diseases. This contrasts with the sterile soil conditions that are maintained in most other greenhouses.

Click beetle and its larva the wireworm (Limonius agonus)

An earwig (Forficula auricularia) caring for her eggs.

Earthworms aerating and fertilizing the soil.

Ground beetle (Calsoma scrutator)

Rove beetle (Staphylinidae)

Sowbugs

Nematodes, actomycetes, molds, bacteria, rotifera, protozoa (some visible under a 10x lense, others only under far greater magnification)

Slugs (Mylax gagetes)

Centipede (Chilopoda)

Millipede (Diplopoda)

PEST: Aphids (Myzus persicae), to about 1/8". Sucking leaf juice from nasturtium; giving birth to live young; growing wings to fly to less populated area.

BENEFICIAL: Ladybugs (Hippodamia convergens). Larvae hatching from a cluster of orange eggs, 1/8"; pupae; larvae (up to 1/2"), eating an aphid; adult, about 1/4". Larva and adult also eat many other pest insects.

PEST: Whiteflies (Trialeurodes vaporariorum), about 1/16". Sucking leaf juice from sweet pepper plant; laying eggs; white larvae.

BENEFICIAL: Encarsia formosa, about 1/30". Lay eggs in whitefly larva, which then turn black.

PEST: Two-spotted spider mite (Tetranychus urticae), about 1/50", sucking leaf juice from tomato plant. Tiny yellow eggs. Mites spin webs when crowded.

BENEFICIAL: Predatory mite (Phytoseiulus persimilis), about 1/25". Tiny white eggs. Adults eat spider mites.

PEST: Leaf miner (Liriomyza), about 1/4", on French sorrel. Adults lay eggs between the upper and lower layers of leaves. Larvae make tunnels, eating as they go.

BENEFICIALS: Trichogramma, Chalcid and Braconid wasps, 1/10-1/8", lay eggs in various stages of leaf miners.

PESTS: Aphids sucking the leaf juice from pea plant.

BENEFICIALS: Flower fly (Syrphidae) 1/3 to 1/2", eggs, pupa, larva.

Brown lacewing (Hemerobiidae), about 1/2".

Green lacewing (Chrysopidae) adult, about 3/4", cluster of eggs elevated on thin stalks, pupa and larvae.

PEST: Thrips (Thripidae), about 1/8", eating chard.

BENEFICIAL: 1/8" hopping bug, identify uncertain.

I sent specimens for identification to the 2 top entomologists in North America. One said it was beneficial, the other said it was a pest. In my experience this bug is beneficial, always appearing within 2 weeks of arrival of thrips, and soon brought them under control.

PEST: on left leaf, European cabbage moth (Pieris rapae), eggs, caterpillar, to about 1.5", pupa.

On right leaf, diamondback moth (Plutella xylostella), pupa and caterpillar, to about 1/2".

BENEFICIALS: Braconid wasp laying egg in moth egg. Chalcid and braconid cocoons attached after having parasitized European cabbage moth caterpillars.

PEST: Cabbage looper (Trichoplusia ni) caterpillar, to 1", and moth on red mustard leaf.

Also Cross-striped cabbageworm (Evergestis rimosalis) caterpillar, to about 1", shingled eggs, pupa and moth.

BENEFICIAL: Braconid wasp laying its egg to parasitize a moth egg.

PEST: Aphids on lettuce leaves.

BENEFICIALS: left, mummified aphid was parasitized by chalcid wasp. Empty aphid mummy vacated by mature chalcid wasp. Right, gall midge (Aphidoletes aphidimyza) adults, with caterpillars, to about 1/8", sucking blood from aphids.

The Solviva Greenhouse is Heated by Animal Body Heat in Addition to the Direct and Stored Solar Heat.

One hundred chickens and thirty rabbits can live happily in separated areas in the darker areas of the greenhouse. Each animal puts out 8 BTUs per hour per pound of bodyweight, amounting to roughly the heat equivalent of 2.5 gallons of heating fuel per animal per 6 cold months. Their warm composting bedding also puts out considerable amounts of heat.

Equally important, the animals and their bedding enriches the air with co2, which greatly increases the productivity of the plants.

Some of the chickens happily munching on fresh greens in mid-winter. The bedding builds up to 18 inches over a whole year, with weekly additions of old leaves. This becomes 300 cubic feet of superb compost which then is transformed into superb quality organic produce.

This picture also shows one of the waterwalls which serves the dual purpose of storing solar heat and preventing the chickens from entering the garden room.

Contrary to most farms, the Solviva chicken yard is not a muddy mess. Instead, species of grass have self-evolved here that stay productive even with chickens scratching and pecking for worms.

Donna ready to deliver the marvellous Solviva eggs. Due to the excellent air quality, fresh daily greens and freedom of movement that the chickens enjoy, these eggs are 30 percent lower in calories and cholesterol than normal eggs, and far higher in vitamins.

This shows one of the spacious multi-level communal dens for the female and adolescent angora rabbits. These dens are tucked away out of the sun under counters and along the waterwalls.

Two male rabbit friends sharing a den separate from the females.

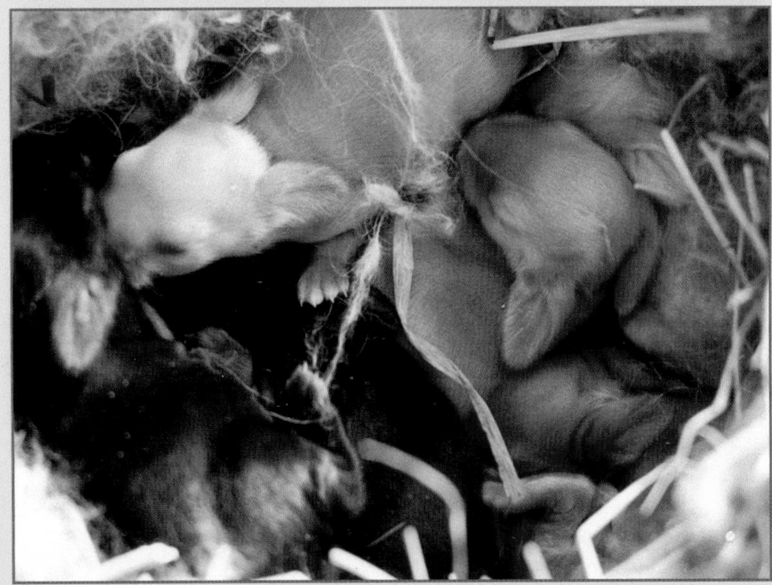

Eight adorable 6-day-old baby bunnies snuggling deep down in a cozy angora-lined nest.

Ruthie feeding hay and grain to the sheep .

Love Dove, Snowbelle, Honeydew, Heather, Miss Silver Velvet, and little Angel Pie by the sheep barn flanking the north side of the greenhouse.

Katherine Rose

Annika cuddling little Cirrus.

Rafe Brown

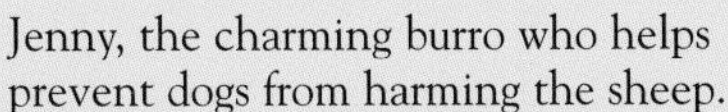

Jenny, the charming burro who helps prevent dogs from harming the sheep.

Ruthie Dreier

Wool-gathering under the old apple tree for the annual feast.

Ruthie Dreier

Katherine Rose

Katherine Rose

Katherine Rose

School children were enchanted when visiting the greenhouse, and quickly understood the implications for the future. Their minds started churning with ideas for solarizing homes, schools and cars, protecting dolphins, frogs and birds. Teachers and parents had never before seen children so motivated to do research and produce reports.

Katherine Rose

Imagine solargreening schools, improving education, air quality, health, while reducing heating bills, without adding to the cooling load. This proposal is for an elementary school in Manhattan.

PROPOSALS FOR A LIVABLE FUTURE

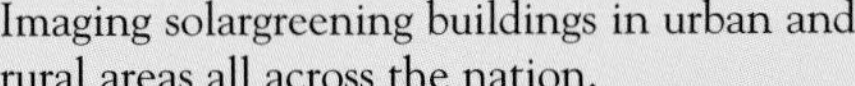

Imaging solargreening buildings in urban and rural areas all across the nation.

The concept of solar design is so poorly understood that any house with a few south-facing windows and skylights is often referred to as a solar house.

A highly effective solargreen addition can be added to the south wall and roof of an existing house, or even to the east or west ends.

A fullfledged Solviva Solargreen home, here seen from the southeast. The whole south roof is solar absorbing, with the solar heat circulated to storage in the slab-on-grade foundation, or under a full basement, thus reducing the heating bills, including hot water, by about 80 percent. In warm weather the hot water is still being solar-heated, but all the solar-heated air is vented out through the ridge, keeping the house nice and cool.

The small photovoltaic array in front is enough to power the essentials during power-outages.

The same house seen from the southwest. This home also has a Solviva compostoilet in the bathroom inside the low green door. The toilet barrel is removed periodically and set to compost in the small adjacent shed.

A Solviva Solargreen building could be built in almost any style or choice of material.

Southwest adobe style.

A design for a **One-acre Solviva Farm**,
capable of producing a gross income of more than $500,000 per year,
without any heating fuel, cooling fans or toxic chemicals.

The energy-independent **Solviva Backyard Food Factory**, enough to provide vegetables, eggs and meat for a family.
Ventilation, water and food can be automatically regulated to enable absences of a week or more.

Painting by Kirsten Edey

Imagine a home with a thriving garden right in the kitchen, while winter rages outside.

Imagine this in a highrise apartment in New York City
(excuse my crude rendition of the cityscape I superimposed outside the windows on Kirsten's beautiful painting.)

It could even be done to a huge building such as this one in New York City.

A Solviva solargreen 4-classrooms school addition.

MORE PROPOSALS FOR SOLARGREEN LIVING

A retrofit to a corporate building in Minnesota.

A retrofit to a health spa in Minnesota.

A large Solviva greenhouse addition to the east side of a large home in Minnesota.

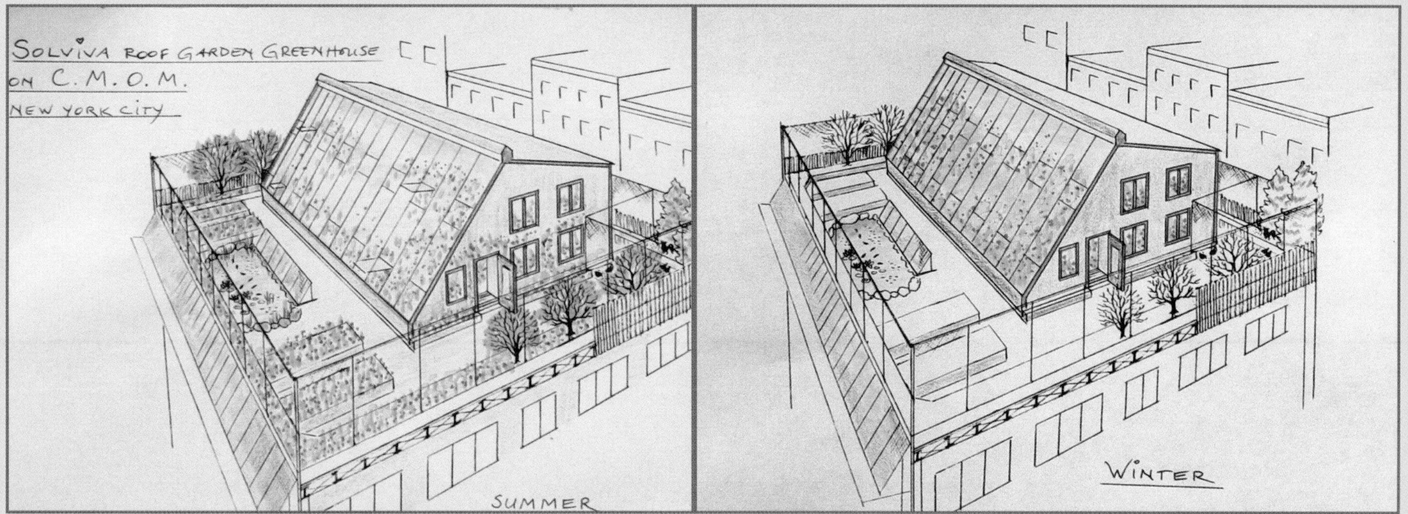

A rooftop Solviva greenhouse for the Children's Museum of Manhattan (CMOM) in the center of New York city.

A restaurant/business center with a mini-farm, on two-thirds of an acre.

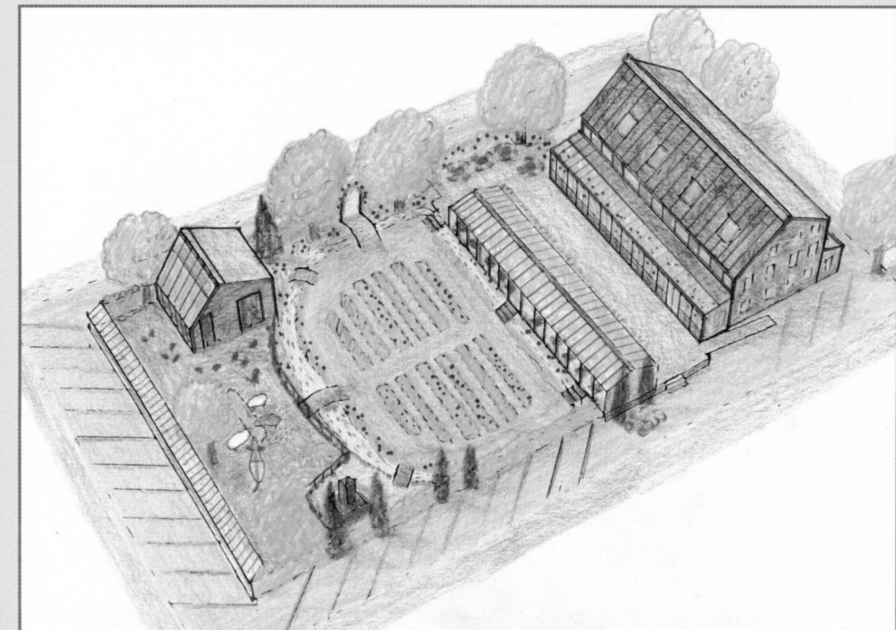

The Solargreen White House.

THERE IS SEVERE WATER POLLUTION ON MARTHA'S VINEYARD, AND IN MOST OTHER COMMUNITIES IN THE UNITED STATES AND THE REST OF THE WORLD.

Many ponds, lagoons, harbors, estuaries and marshes exhibit serious algae pollution. Most people don't notice this as they drive by. Close up you can see, smell and feel the evidence, but that it is only the tip of the iceberg. It is far worse under the surface where fish and shellfish, and the whole ecosystem, are choked by thick masses of algae. The individual algae organisms have very short lifespans and as they die they rot, which robs the oxygen out of the water. This algae pollution has all along been blamed on ducks and geese, on farm animals and lawn fertilizers, and on road runoff. But on the Vineyard and most other places this constitutes less than 10% of the problem. **At least 90% of this algae pollution is caused by vast amounts of nitrogen flowing with the groundwater from thousands of septic systems.** It matters not at all whether these septic systems are 100 feet or 100,000 feet away.

This closing was due to contamination from the Edgartown sewage treatment facility, local septic systems and leachates from the nearby dump.

Tidal pond next to Martha's Vineyard Hospital, which feeds into the Lagoon beyond. The shellfish beds are often closed due to pollution.

Massive algae infestations along outer Vineyard Haven harbor, caused by normal functioning septic systems.

James Pond, adjacent to Lamberts Cove beach, once a pristine and productive shellfishing area and a great place for small children to swim, now foul and unproductive. **Cause:** primarily the nitrogen flowing from hundreds of septic systems from a zone of influence of several square miles.

Sunset Lake, which flows into the Oak Bluffs harbor and its shellfish beds.

A tidal marsh, such as this one by **Farm Pond in Oak Bluffs**, is an excellent absorber of excess nitrogen. But this is what happens when the marsh is overwhelmed by too much nitrogen: thick algae masses have smothered and killed large areas of the marsh.

Most of this pollution is caused by the standard methods of managing wastewater, which are not only allowed but are indeed required by the Department of Environmental Protection (DEP) in Massachusetts and other states across the nation.

These regulations cause not only terrible pollution but are also very costly and require the removal of valuable trees and other landscaping.

Here, under construction for a restaurant on Martha's Vineyard, is a typical example of standard septic system. Several 12-foot-deep trenches were dug, then filled with gravel and concrete galleries for the purpose of dispersing the effluent from the septic tank.

Not only is this septic system very costly, but it is also incapable of sufficiently reducing the high levels of nitrogen (from human body waste) that is contained in the wastewater. Thus, large amounts of nitrogen leach into the groundwater, and then travels with it to the nearest open water, in this case Lake Tashmoo. One could hardly devise a more effective way to get the largest amounts of nitrogen pollution as quickly as possible to open water.

A similar costly, polluting septic system for a home in Oak Bluffs, which required the destruction of dozens of trees.

A costly raised septic system impaired the beauty of this small cottage in Oak Bluffs, but the nitrogen reduction in the system is minor or nil.

Twenty-two beloved shade trees were removed to make room for the new septic system that was required before this small home could be sold. The system cost $10,000, not counting several thousand dollars more for extensive landscape restoration. This is a typical example of economic hardship imposed on many with small income.

These regulations would be understandable if the new septic system substantially reduced pollution, but it is quite the contrary. In actuality, some 400-500% more nitrogen leaches into the groundwater than from the old septic system. This is because the nutrients, grease and particulates flowing into the old leaching pit over the years develop into a water-tight biomat liner. This biomat causes the leachwater to disperse closer to the root zone of the surrounding plants, which absorb a large portion of the nitrogen, and indeed benefit from it. The new septic systems, with greatly increased leaching areas, will take many years to develop an effective biomat.

This is a typical central sewage treatment facility. It serves only a small part of the small village of Edgartown, and cost an appalling 13 million dollars. Hundreds of trees were destroyed for this installation, trees which could have consumed and benefited from all the nitrogen, at a fraction of the cost. Instead, this facility releases thousands of pounds of nitrogen per year into Edgartown Great Pond, which periodically causes massive shellfish kills. . . and it smells bad.

> *One of the reasons why high cost projects such as central sewering are continuing to be promulgated, and the far less costly innovative waste water management systems continue to be suppressed, is that consulting engineers get paid a percentage of the total cost of the project. Thus, even though the percentage is somewhat smaller for a higher cost project, a $10 million project will generate a design fee of $640,000 while a $2 million project will generate a fee of only $150,000. In addition, it is more work for the engineers to design a system they have no prior experience with, far less work to design a system the same as many others they have already done.*
>
> From GAO (General Accounting Office)

All these concrete leaching chambers and gravel, and $100,000, were required for dispersing the runoff into the groundwater from the parking lot of a local supermarket. Again, this design constitutes the fastest way to get maximum pollution into the nearest open body of water.

Here is a far better solution, in Denmark: the whole parking lot has paving blocks interspersed with top soil and low-growing grasses which clean the pollution before it ever reaches the groundwater.

In 1977 I discovered something that hardly anyone else was aware of: urine, diluted about 1:10, is an excellent fertilizer. Witness the Indian Poke plant which grew over 10 feet tall, twice as tall as usual.

Here is one square yard of garden area planted with 270 onions, and fertilized only with diluted urine.

This golden elixir received many new names: Peace-on-Earth, You're-in-charge, You're-in-power...

It produced 270 large-size onions which lasted until the next summer's crop came in.

For many years I thought we all had to convert to compost toilets to protect our water quality. Here, between the bathtub and the sink, is an example of the system I designed, very low-cost, odor-free and easy to manage, and requiring no basement. An exterior view can be seen in the southwest view of one of the house designs a few pages back.

But over the years I realized that no matter how wonderful and convenient compost toilets can be, most people will continue for a long time to be strongly opposed to choosing that option. So in 1985 I decided to design a composting system that uses a flush toilet. I removed my upstairs compostoilet and replaced it with a regular toilet. It flushes into the insulated composting chamber (pictured open here), into which I put 12 cubic feet of Biocarbon Filter medium (mostly old leaves) and 3000 earthworms.

The amazing result is that this chamber, with a volume capacity of 25 cubic feet, has over the last 25 months received about 2500 flushings (from 1-6 people per day), 140 rolls of toilet paper and about 35 cubic feet of Biocarbon Filter medium. This has all been reduced to about 9 cubic feet of superb earthworm castings, the best of all soil amendments.

This is clearly Earthworm Heaven, and they have multiplied to millions. It has never caused any smells or flies. Everything flushed into the chamber is digested in about 3 days in summer, 5 days in winter, without any heating device. I never would have dreamed that this was possible, until I proved it.

The effluent drains from the composting chamber into a series of prolific Greenfilter flower beds (compare with the adjacent vegetation browned by drought conditions). The whole system reduces nitrogen, COD and BOD by some 90%.

One of the by-products from the Solviva Biocarbon Wastewater and Flush Toilet Purification Filter systems: superb quality compost.

One of Man's best friends: the Earthworm.

Ruthie Dreier

The Solviva graywater garden: this area, with its thriving roses, dogwood, pines, spruce and grasses has received all graywater since my home was completed in 1981. Over the past 17 years these plants have successfully processed over 500 pounds of regular detergents, shampoos and cleaners, and 45 gallons of chlorine bleach.

The flower garden is thriving with the compost from the toilets.

A long-standing hot topic on the Vineyard is: what to do with septage (pump-outs from thousands of septic tanks) and sludge (from the Edgartown sewage treatment facility). Currently Oak Bluffs is shipping some 400,000 gallons of septage off-island for treatment, at a total cost of 17 cents per gallon. Some businesses are paying $30-55,000 per year for tank pumpout. The town is considering paying six to eight million dollars for a local septage treatment facility.

Last winter I ran a small Biocarbon septage treatment experiment in my unheated garage. I poured septage through a 4-stage Biocarbon Filter system in which the first stage captured the solids and the next 3 stages filtered the effluent through Biocarbon Filter medium. After a total filter passage time of 5 minutes the effluent was odor-free, and the nitrogen, BOD and COD reduced by 90 percent.

The sludge was emptied into a bed of Biocarbon Filter medium and covered with same, which immediately eliminated all odors. It was processed by earthworms and microorganisms into excellent compost in less than one month.

Grass thriving in the Greenfilter, the final stage of the Biocarbon filter.

Healthy Indian poke and tomatoes (which produced dozens of fruit) growing in the Biocarbon Brownfilter and septage effluent.

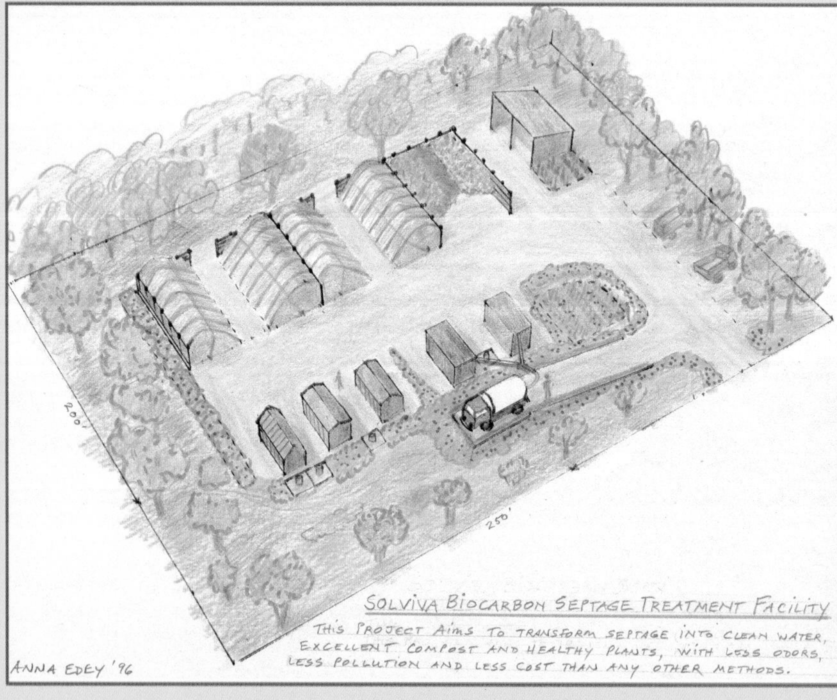

Based on this experience I designed an upscaled **SOLVIVA BIOCARBON SEPTAGE TREATMENT FACILITY,** requiring **only one acre of land**, capable of transforming 5 million gallons of septage per year into clean water, excellent compost and healthy plants, without any odors, at a fraction of the cost of conventional systems.

A GALLERY of some of the other PROPOSALS for SOLVIVA BIOCARBON WASTEWATER PURIFICATION FILTERS for MARTHA's VINEYARD, 50-70% less costly than other systems.

A home system capable of processing up to 440 gallons of wastewater per days (note small 4-by-4-foot shed attached to the front corner of the home: this contains the Brownfilter portion of the Biocarbon Filter).

A larger system capable of processing 3000 gallons per day.

An early design for the Black Dog Tavern in Vineyard Haven, generating nearly 5000 gallons per day.

Public restrooms for Kennebec Ave, Oak Bluffs

Small building containing 3 standard and 2 handicapped-access restrooms, with Solviva Biocarbon composting toilets, for the Gay Head beach (were there is no water available for flush toilets)

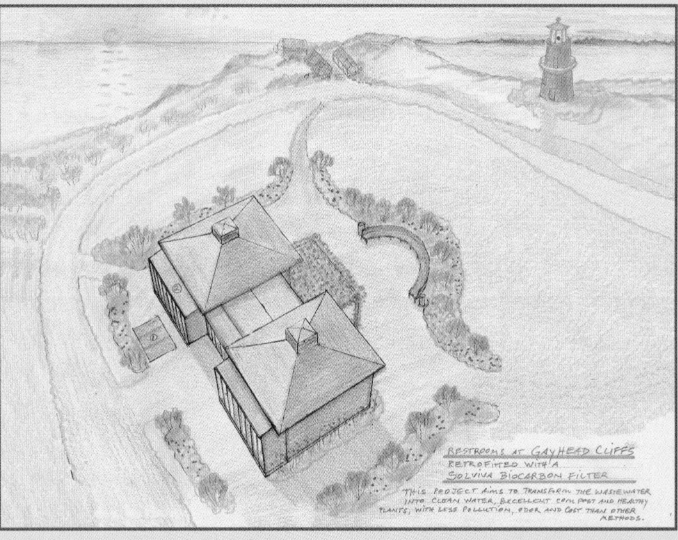

Public restrooms for the Circle at Gay Head Cliffs.

HOW SHALL WE LIVE?
IT IS NOW SELF-EVIDENT THAT WE CAN AND MUST CHOOSE.

Shall we choose to continue growing food, managing solid wastes and wastewater, providing electricity, transportation, heating and cooling with the current normal methods that cause severe pollution and depletion of our resources, and threaten our personal, community and national health, economy and security.

Or shall we choose to live, in ways that are sustainable, bio-benign, solar-dynamic, that protect our environment, resources and economy, and that foster good health, peace and security.

THE CHOICE IS UP TO US, THE PEOPLE OF THIS AND OTHER NATIONS ON PLANET EARTH.

PART TWO

A VISIT TO SOLVIVA

❦ ❦ ❦

Winter in My Home

Two months later, in the midst of a record-cold and dark winter, the temperature is well below zero degrees F, and the howling wind creates a fierce windchill factor. The landscape is deep in snow, and my car barely makes it up the driveway past the Solviva Winter Garden greenhouse. This greenhouse has never needed any supplementary heat since its completion in 1984. It has indeed stayed warm enough to continuously produce bumper crops with only stored solar heat and the little resident heaters, the 100 chickens and 30 angora rabbits. Today, however, is the coldest it has ever been since the greenhouse was built, and I am worried. I am tempted to stop and check in, but I know how it is just to "check in": once inside, it is impossible to leave because of the pure wonder of this garden in winter, and also because there is always some task that begs to be done. But today is my day off and I have other plans, so I continue up the driveway, slipping and sliding up the hill to my garage.

Bundled up in full winter gear, with an armload of groceries, I struggle against the arctic blast, down the path toward my home. Before going inside I pull out a bag of birdseed and trudge around the corner through waist-high snowdrifts to fill up the bird feeders. Biting wind snarls in under my muffler, and my fingertips ache from the intense cold as I fumble with the mechanics of the feeders. A little chickadee hops right up, looks me straight in the eye and says "many-thanks-to-you, many-thanks-to-you". As our souls meet, a profound joy courses through me. With squinting eyes I inhale the glistening white landscape etched with naked trees and bushes and their blue shadows, the deep blue sky, the brilliant sun.

However, this is not the time or place to linger, for the tips of my fingers and nose are screaming for me to seek shelter. So I retrace my deep footprints around the corner, open the front door, and quickly step inside. In the entry hall I peel off my outer layers and then open the inner door to the living room.

With relief I deeply breathe in the fragrant warmth of my home and the beauty of the sun pouring in through climbing and cascading green leaves and red, orange and yellow flowers. Even after several years of living in this house, I still find it hard to believe that this is possible, that the solar energy radiating in through the south glazing is enough to keep my home cozy even on such a cold day.

Enchanting scents waft from jasmine and honeysuckle, sweet peas, pineapple sage and peppermint geranium, orange and lemon blossoms, and compost-rich living earth. Some of the branches and vines reach into the living room and kitchen, the weaving studio, and even upstairs. This garden room has no walls or doors separating it from the living areas but is instead fully integrated with the whole house.

I reach up to pick one of the many juicy tomatoes hanging from numerous 30-foot vines suspended under the skylight that runs the full length of the lower 30 inches of the main roof. These tomatoes are even sweeter than the finest summer tomato, perhaps because they develop more sugar as they ripen more slowly in winter. These same six tomato plants have been producing continuously for four years.

I take two steps down into the garden, then across the warm concrete floor and along the steppingstones set among the carpeting of baby-tears, to the old clawfoot bathtub in the west corner. It is surrounded by a bird-of-paradise plant with five flower clusters in full plumage, an enormous Monstera deliciosa with seven ripening fruits, a large hibiscus bush with a dozen red blossoms, and an arbor of heavenly scented flowering ginger. Next to the tub is the salad garden, pouring forth a continuous cornucopia of succulent greens and herbs.

I turn on the hot water, sprinkle in my favorite bath salts and oils, and pick a handful of peppermint geranium, rosemary and sage, tossing them in too. Then I turn on the hose to self-water the southeast quadrant of the garden. I peel off most of my clothes in response to the solar warmth and putter around a bit, snipping a withered flower here, twisting up a vine there, picking a bouquet of flowers and ferns for the kitchen table.

Next I harvest some 10 different varieties of red and green lettuces, plus arugula, mustards, kale, lemony French sorrel, red-veined chard, radicchio, mizuna, tah tsai, watercress, parsley, dill, fennel, and crunchy radishes. I give them a quick rinse and shake and seal them into a plastic bag and put them in the refrigerator. And I take one perfect brown egg laid by the greenhouse hens and set it to hard-boil. As I stand in the kitchen, close to one of the warm-air ducts, I can just faintly hear the hum of one of the fans that move warm air from the solar roof into storage in the insulated foundation.

Before the tub is full, I have time for one more task: pollinating the tomato flowers. For a few minutes every two days I become a fairy godmother, touching each fully open tomato flower with my magic wand (a watercolor brush taped to an extension stick), ping... ping... ping... A day or two later the flowers wither, but instead of falling off, the way they would if not pollinated, the base of each flower will fatten into a tiny green tomato, grow larger and larger, turn yellow, then orange, then red, shiny, sweet and juicy.

When the tub is full and I have gathered my phone, tape recorder and a cup of peppermint tea to be within easy reach, I get in and slowly sink into bliss. Mozart's C-major piano concerto joins with the birds singing and lunching at the feeder right outside the window.

My body is afloat, slowly rising and falling as I inhale and exhale deeply. My mind is free as I release tensions from toes and fingers to neck and brow. I enter a state of peace and fulfillment.

(...One of my golden memories is being in this tub one blizzardy Christmas with my two little grandchildren, blowing bubbles and splashing water all around....)

I finish off with a shower, rainbows arching through the mist, diamond water droplets refreshing the surrounding plants. I leave the hot water in the tub to slowly give off its warmth during the upcoming cold night and then dry off with a sun-warmed towel. I step up to the kitchen and take the bag of salad out of the refrigerator, put a handful of crisp greens on a plate, top it off with my favorite dressing and a few tasty nasturtium, borage and fennel flowers, and then step back into the garden and sink into the lounge chair. Here I bask in comfort, just a couple of feet away from the cold raging on the other side of the window panes - and the only furnace that is on is the big old nuclear power plant up there some 95 million miles away.

What a blessing it is to be munching on these greens that grew from tiny seeds right in my home, in a harmonious blend of water, soil, compost and sunlight, without any toxic substances whatsoever. They now release a symphony of exquisite flavors and health-enhancing vitamins, minerals, enzymes and active vibrant life force. The egg is far higher in vitamins and 30 percent lower in calories and cholesterol than factory eggs, because the lucky hens live in freedom in a clean environment and receive fresh greens every day.

People tend to think I must spend half my time just caring for this indoor home garden. The fact is that it takes less than an hour a week, and considerably less than that when I grow just ornamentals. Two or three times a year I spend a few hours pruning back and repotting some of the plants and enriching the soil with compost and rock powders.

People ask, Don't insects fly and crawl all over the house? No, insects are smart, they much prefer the garden area. However, a few shy, slender daddy-longlegs-type spiders roam discreetly around the house and do me a great favor by controlling the wool moths which would otherwise devastate the wool yarn and weavings in my studio. These spiders do nothing worse than leave webs in the ceiling and a few shriveled husks of insects that they have captured and eaten. I doubt the living quarters of my home have many more insects than any other country house with wood beams and paneling.

And people ask, Doesn't mildew form everywhere because of the humidity from the indoor garden? No, because the air circulation in the whole house is excellent due to the solar-heating system and the strategic placement of ducts and vents that move warm air down and around. Also, when the woodstove is lighted, it acts as both an air circulator and dehumidifier. The extra humidity in the house in winter is actually a great benefit. In normal buildings the air is too dry during the heating season, which can lead to respiratory problems and illness, dry skin and cracked lips, static electricity, cracks in furniture and floors. People with asthma and allergies comment with relief that they can breathe more freely in my house. People's eyeglasses do fog up for a minute when they come in from the cold, but that is not much of a problem.

Just before sunset I insert the movable insulation panels that fit snugly into the window frames to keep the solar warmth in the house. These panels spend the day in a long built-in pocket below the greenhouse windows. It takes less than a minute to put them up, and the same to take them down in the morning. On the coldest, darkest days I leave most of them up in order to keep in the warmth.

Because of the brutal conditions outside I make a fire in the evening. I keep a stack of old newspaper by the Franklin stove and first make a bed of thin twists of paper for kindling, followed by increasingly thick logs of rolled paper, without any roller or ties, to fill the whole stove. I wedge them in crosswise to keep them from unfurling and to let air through, and then close the stove doors and leave the flue damper open for about half a minute, until the draft creates a good fire. This whole process takes less than two minutes. Then I close down the draft, and the fire continues burning hot for a couple of hours.

The stove sits in an alcove of brick and concrete, which picks up so much heat that it is still warm the following morning. A 100-foot coil of copper pipe, connected to the water preheating tank, lies on top of the stove. This can heat the 80 gallons of water to 130 degrees F in just a few hours, which is then complemented by the standard water-heating tank as needed. Thus my space heating bill is zero, while the water-heating bill is reduced by some 80 percent.

I know that a little tree frog, only about an inch long, is spending the winter in my home, because now and then it peeps a few times loud and clear. Many times I have tried to sneak up ever so carefully to see it, but I swear it can hear my heartbeat, for it stops peeping as I approach, and when it is quiet it is impossible to find it in among the leaves. This evening, as I sit reading under the lamp next to the garden, I have the distinct feeling that someone is watching me from behind. I slowly turn my head, and there on a branch with a bright red hibiscus flower, about 12 inches from my face, sits my little housemate, looking me straight in the eye.

Winter in the Solviva Greenhouse

That night is the coldest it has been in decades, and extremely windy. I sleep fitfully, concerned about the greenhouse: Can it possibly survive this night without backup heat?

At 4 a.m. I wake with a start as ice and snow come crashing off the roof and the gale rattles my windows. Now I am wide awake and really worried. Rather than lie there fretting, I get up and pull layers and layers over my pajamas, push through a 5-foot snowdrift right outside the door, and set out across the fields. It is an 800-foot passage. The surface of the snow sometimes supports my weight; other times I crash down above my knees. One false move and I could break my leg and be trapped in the snow with no one to hear my cries for help except the brilliant moon and the silver-edged clouds chasing matching black shadows across the landscape.

Twenty minutes later I approach the greenhouse, nestled in a snowdrift at the far end of the pasture. Whirlwinds of white wisps whip around in the moonlight. My breath has turned to ice on the muffler pulled over my face. I hastily shovel away several feet of snow blocking the west entrance door, wrench the door open and quickly close it behind me.

To my utter surprise, in here it is like a balmy night in June. The thermometer reads 55 degrees Fahrenheit. The 30 angora rabbits that help warm the greenhouse with their body heat are quietly muffling about in their communal dens. I step into the greenhouse, through the jungle of tomato vines, and here the thermometer reads 45 degrees F.

I proceed toward the east end, inhaling the humid, mild air, fragrant with tomatoes and nasturtium, thyme and sage, and living earth. At the far end I step in among the 100 roosting chickens who acknowledge me with sleepy murmurs, cozy in their warm, spacious quarters. The thermometer reads 70 degrees F and this warmth is generated by the body heat of the chickens.

The sheep, enclosed in the barn along the back of the greenhouse, their bales of hay stacked up against the wall, further help protect the greenhouse on this blizzard night.

Thus, while the outside temperature is 5 degrees below zero F - though actually much colder because of the windchill factor inside the Solviva Winter Garden greenhouse it is warm enough to maintain a thriving garden, abundant with vegetables and flowers, without any heating fuel. I can go back to bed without worrying about the greenhouse freezing. So I tromp back across the fields, a bit more easily now as I retrace the deep footprints I left on the way down, feeling entirely at peace and as one with Earth, Universe and self. This is true plenty, freedom and security.

The next day it is still extremely cold and windy, with brilliant sunshine. The min/max thermometer shows that during the night the greenhouse never dipped below 43 degrees F. By 9 a.m. it is 75 degrees inside, and I turn on the hose and with quick quivering motions provide everything with a light refreshing shower.

Two fans, powered by the sun shining on the photovoltaic panels, hum as they force hot air from the top of the greenhouse down through ducts and into heat-absorbing water-mass storage. Some of the heat-activated vents are slowly opening, increasing air circulation and preventing overheating. The massive waterwalls are passively absorbing the solar heat.

Twenty-five varieties of lush greens and herbs fill the raised beds, with names like hon tsai tai, arugula, tah tsai, Osaka mustard, mache, radicchio Treviso, mizuna, and the divine lemon-flavored sorrel de Belleville. Above them rise 150 growtubes hanging in seven tiers to the top of the greenhouse. They are overflowing with 25 varieties of lettuces with names as lush as their appearance and flavor: Lollo Rossa, Rouge Grenobloise, Rosalita, Merveille de Quatre Saison.

Three of the seven tiers of growtubes are set upstairs along the catwalk, and here are also a steady succession of dozens of seedling flats ranging from just seeded to 3 inches tall and ready to be planted into the raised beds and growtubes. Here they sprout and grow strong without any extra light or warmth, even through prolonged cloudy cold spells.

Hundreds of tomatoes are ripening on 15-foot climbing and cascading vines. Along the north wall, where the light is too dim for greens to thrive, there is a tall wall of nasturtiums with thousands of blossoms in infinite varieties of pastel and deep velvety colors.

Fennel reaches 8 feet, tipped with 6-inch umbrels of tiny yellow flowers, exquisitely anise-flavored. The delicate red trumpet flowers of the 6-foot pineapple sage bush yield little drops of nectar that actually taste like pineapple. Another variety of scented sage reaches 16 feet tall, covered with sweet pink flowers. A lime geranium yields exquisite fragrance, as do carpets of honey-flavored sweet alyssum.

A few square feet of bed was planted with radish seeds three weeks ago and now yields 1-inch red and white globes, mild and succulent. Another patch contains hundreds of the sweetest carrots. I pull up one daikon and find to my amazement that it has a three-pronged root, each prong pure white, thick as my wrist and 16 inches long.

Many branches of one pumpkin plant (a volunteer from the compost) cavort 15 feet in all directions, with pumpkins up to 12 inches supported and hung on various improvised shelves and slings.

The breath from the chickens and rabbits and their bedding enriches the air with several times the normal level of invisible molecules of carbon dioxide. The plants breathe in the CO_2 through the stomata on the surface of their leaves, and the CO_2 enrichment causes them to grow much faster and healthier because it provides them with more carbon building blocks to create plant tissue.

Five dozen eggs per day and the angora wool more than pay for the animals' feed, while their body heat, CO_2, compost fertilizer and good company are free fringe benefits.

Ladybugs eat aphids, miniscule Encarsia formosa wasps lay eggs in the pupae of whiteflies, while green lacewings flit about like little fairies in search of any vegetarian insect. Syrphid flies seek nectar from fennel flowers, hovering like hummingbirds, and a dignified praying mantis is surveying the scene and, blessing me with her eye contact, pronounces it good.

That day, in spite of the cloudy, cold, short days of mid-winter, we harvest, picking leaf by gorgeous leaf, wash and bag 80 pounds of Solviva Salad, enough for 1,280 servings. Some go off by UPS truck to the finest restaurants in the Boston area, and the rest is delivered to customers in local Vineyard restaurants and stores.

Summer in the Solviva Greenhouse

Seven months later we are in the middle of a sweltering record-hot summer. There has been hardly a drop of rain for three months. This day in August is more of the same. By now, the lettuces in most other gardens have bolted. But because of the Solviva growing techniques, the outside garden is a continuously productive patchwork quilt of lettuces and other salad greens in brilliant rosy reds, deep wine reds, lime greens and sun greens, dark greens and blue greens.

Inside the greenhouse it is surprisingly cool, and yet there are none of the expensive, roaring, energy-consuming exhaust fans that standard greenhouses run continuously. The hot air is rushing out through the top vents, and cooler replacement air comes in through bottom vents, and east, west and north doors. Surprisingly, the greenhouse is as stunningly productive in summer as it is in winter. It is hard to stay away from superlative adjectives when describing this scene.

At the west end 10 varieties of peppers grow up to 6 feet tall, yielding hundreds of fruits from sweetest to hottest.

The center of the greenhouse is filled primarily with 10 different varieties of melons and cantaloupes, with vines over 30 feet long loaded with ripening fruit suspended in net bags and slings. Their flavor surpasses anything grown in an outside garden.

A tall bower of European cucumbers fills the east end, with 30-foot vines and 16-inch leaves, and foot-long tendrils seeking the next handhold. Drooping from this bower are 18-inch delicacies, almost 2 pounds, with tender thin skin and no seeds.

Tomato plants with 30-foot vines form other bowers, heavily laden with thousands of ripening fruit. The wall of nasturtiums continues to flourish along the north wall, now cool and shaded from the high summer sun.

In front of all the climbing vines is a long bank of five different kinds of basil, more tender, productive and flavorful than any grown outdoors. In here are also many different greens of the Crucifera family. When planted in the outside garden, these greens are devastated by flea beetles, but for some reason these pests are not in the greenhouse.

Bees, hover flies and lacewings flit around sipping nectar and at the same time performing the important task of pollinating the flowers. Without this service there would be no fruit, unless we take the time to touch each of the hundreds of flowers every two days, which we need to do in the winter.

A quick shower for the whole greenhouse a couple of times daily provides highly effective evaporative cooling. As in winter, there are no mildew problems. Most plants are more productive, tender and flavorful in the greenhouse in summer than they are outside. In here they are sheltered from the harsher conditions that prevail in the outdoor gardens, such as whipping winds and occasional pelting rain, and the full blast of the sun's ultraviolet radiation.

I walk across the fields toward my home, passing the grazing sheep and Jenny the burro. They are followed by the chickens who happily clean up any parasite eggs or larvae, thereby gaining good protein while keeping the sheep and burro healthier. Grazing, they provide ongoing mowing without which the fields would become an impenetrable tangle of poison ivy and brambles. In the process all parts of the entire ecosystem, from plants to insects to animals to humans function harmoniously and effectively together.

Summer in My Home

Even on this very hot day my home is delightfully cool, without any exhaust fans. The hotter the sun shines on the solar roof, the faster the hot air escapes through the top vents, pulling house air out with it. The water pipes incorporated in the solar roof provide for most of the hot water needs. Refreshing breezes enter through doors and windows, strategically placed to scoop in the prevailing winds. The plants in the greenroom are thriving in full sun, even cool-loving plants such as broccoli. The relative coolness of my home on a hot sunny day is a great surprise to people who think of solar houses as uncomfortably hot in the summertime.

I take a shower outside and the water runs down a lined trench to irrigate a Viburnum odorata, Japanese maple and other moisture-loving plants that grow around the deck. I happily recall taking hot showers here in mid-winter. Standing in the snow under a hot shower is like drinking hot chocolate with cold whipped cream.

Summary of Advantages of Living the Solviva Way

I have found many advantages of living the Solviva way. First of all, there is the matter of money. My living expenses are several thousand dollars less per year than the average home. For one, my solargreen home saves me about $1,200 a year in heating oil costs. A friend in southern New Hampshire could save most of the $5,400 a year she spends heating her home with electricity. Furthermore, I save a couple of hundred dollars on water heating because the water is heated mainly by the sun and free wastepaper fuel, and I save another $75 because by burning the fires very hot the chimney never needs cleaning.

I also save significant amounts on my electric bill because my refrigerator is a whisper-quiet Danish Vestfrost, which uses about 75 percent less electricity than a standard refrigerator, and my light bulbs are compact fluorescents which also use 75 percent less electricity while providing the same amount of light as normal incandescent bulbs.

In addition, I have a Solviva waterless compostoilet, and a regular water toilet that flushes into a Biocarbon composting sewage purification septic system, as well as a Biocarbon graywater purification filter, and these save on wastewater management because I need no periodic septic tank pump-outs ($480 per pump-out in some communities), nor do I have the enormous expense ($8,000-$30,000) of upgrading a substandard septic system or replacing a failed one. The average life span of a standard septic systems is only 10 to 20 years. By contrast, I estimate the life span of a good Solviva Biocarbon septic system to be "forever".

I also save about $100 a year on solid waste management because I compost all food wastes, burn all low-grade wastepaper, and recycle all glass, cans, magazines, corrugated and plastic. Only about 10 percent of my solid waste is trash. The recycling takes no extra time whatsoever. In fact, I save time as I need take only a few trips a year to the dump, because there is nothing in my wastes that causes odors or attracts flies, rats or dogs.

Furthermore, I can save hundreds of dollars on food because of what I can produce right in my home, and because of the superior quality of this food as well as the excellent air quality produced by all the indoor plants, I hardly ever get sick. This saves the cost of over-the-counter and prescription drugs and saves time that would be lost during sickness.

Thus my home clearly reduces my cost of living substantially. It also, even in its state of incomplete conversion to a fully solargreen home (my home is still electrified with oil and nuclear power, instead of solar photovoltaic panels), it causes some 80 percent less pollution than a standard home. For instance, it causes some 30,000 pounds less CO_2 pollution because the heating, cooling and electrical systems require some 1,500 gallons less oil than the average home, and it causes some 90 to 100 percent less groundwater pollution because of the various Solviva wastewater purification filters.

The Solviva Winter Garden greenhouse on the farm saves me the $6,000 or more that a conventional greenhouse of equivalent production capacity would cost to heat and cool, preventing the depletion of roughly 6,000 gallons of oil and emission of 120,000 pounds of CO_2.

The various Solviva greenhouse and farm designs and management systems prevent the bad odors and flies, as well as the pollution of water, soil and air normally caused by conventional farms.

Thus, not only do the various Solviva designs greatly reduce the cost of living and harm to our resources, environment and other species, but they also promote good health and good feelings.

Based on my accumulated experience, I believe that these various Solviva systems can be adapted to work sustainably in any urban or rural location in any climate on Earth, for countless generations to come.

PART THREE

How I Got on to the Path of Seeking Better Ways to Live & What I Discovered Along the Way

ꕥ ꕥ ꕥ

When I was a child growing up in Sweden, Mother Nature was my best friend. I roamed over hills, through meadows and along pristine stream beds and beaches, intimately learning the names and habits of wildflowers, trees and insects. The forests were deep, misty and fragrant, carpeted with spongy mosses and furnished with giant boulders. Lured by the beauty and mystery, I glanced over my shoulder with pounding heart, fearing/hoping for the appearance of the trolls and fairies I knew so well from storybooks. I knew where the mushrooms would pop up after a rain, which could kill you and which were safe and delicious. I knew when the wild strawberries and blueberries would ripen in the dappled birch glens.

My mother kept houseplants blooming and climbing in every window, and I soon did the same. I had gardening grandmothers, each with her own paradise, one in the city, the other in the country. The one in the city gathered horse "pears" from the streets (there were still plenty of horse-drawn carts), the one in the country got manure from the neighboring farm, and each garden resulted in a glorious succession of flowers, vegetables, fruits and berries. I joined the birds and the wasps in search of the ripest cherries and plums, pears and apples, gooseberries, currants and strawberries: abundance without any toxic sprays.

I grew up with relatively clean air and surroundings, in a society with minimal crime or social strife, poverty or unemployment. Design was sensible, efficient and attractive. There were no billboards. The few cars were small, there were no traffic jams, and you could go anywhere with public transportation.

Thus, I was quite shocked when I came to America in 1957 as a young bride, complete with my looms and other weaving equipment. Looking at the memory album in my mind of my life in New Jersey, New York City and Long Island, I see snapshots of gigantic cars sporting tons of chromed fins and fangs, clogging eight-lane highways lined with billboards telling me I'd be happier if I drank this or smoked that. Beyond, I see acres of parking lots, and suburban housing developments with thousands of identical homes with picture windows looking across the street into the neighbor's identical picture window, not a tree in sight. Live now, pay later.

I see forests of chimneys spewing out thick smoke of all different colors, and oil-slicked, trash-filled rivers and harbors belching foul odors.

I see degrading poverty and despair in one block contrasted with ostentatious wealth in the next.

On television I see fire hoses, clubs and snarling dogs attacking people requesting equal rights to voting, education, seating on buses and lunch counters.

I see the pinkest hot dogs, the yellowest mustard and the greenest relish: food coloring in everything.

I see myself in New York City, nauseated from breathing the air and drinking the water, both fouled with toxic substances.

I see people gardening and farming in dead soil with chemicals from a bag and foul-smelling liquids from bottles marked "poison".

I see my first child being born in 1958, while I fight for my right to be awake at the birth, my husband not allowed to attend, and I see the nurses labeling me uncivilized and unsanitary because I insist on breast-feeding my baby.

I see myself waking up one dawn, while summering in Edgartown, Martha's Vineyard, by a screaming roar coming up the road. I rush into the bedroom that faces the street, where my babies are waking up from fright, just in time to see a sickening cloud billowing in through the open window, filling our lungs, coating our skin: DDT for mosquito control.

I see us taking a walk along the Charles River in Boston, and one of my little daughters trips and slips in up to her knee. She is dripping with unspeakably stinky black ooze as I whisk her up and rush to find a place to clean her off.

I see people building fallout shelters, preparing for The Bomb to fall, furnishing them with food, water and blankets, and guns to prevent anyone else from getting into their shelter.

I see myself buying canned and powdered milk, because fresh milk was contaminated with Strontium 90, fallout from atmospheric testing of nuclear bombs.

For a young woman from a clean and nature-loving country, where peace and justice prevailed, the United States was not a good experience in the late 50s. I had never imagined that human beings could be capable of such carelessness and brutality against self, others and Nature.

Then came the 60s with its powerful revolt against waste, materialism, racism, militarism, sexism, and destruction of our planet. Rachel Carson sounded the alarm about pesticides with Silent Spring. It felt as if things could not get much worse and that the times "they were a-changin'."

Yet evidence indicates that since then, in spite of a tremendous rise in right action and awareness and a great deal of improvements, many social, health, economic, resource and environmental conditions in the United States and around the world have deteriorated faster than they have improved. Alas, the same is true in Sweden. Thus, the current costly, polluting, wasteful conventional systems, and the bureaucracy and regulations that enforce them, have us trapped in what many consider to be a hopeless downward spiral towards unimaginable disasters.

In 1972 our family moved year-round to Martha's Vineyard, in search of a simpler, more harmonious lifestyle. We designed a home and built it with sincere but novice carpenters. We installed a Franklin stove for wood heating and a beautiful old kerosene heater. For backup heating we chose electric baseboard heaters because of low installation costs and promises of cheap, clean and reliable electricity.

Unfortunately, our hopes for a more harmonious life did not survive the stress of building a house. Soon our marriage broke up, and I was a single woman living with my three daughters in an isolated, drafty and poorly built house, facing the problems of a harsh winter.

In late 1973 the first oil embargo hit. The cost of oil skyrocketed and with it the cost of electricity. Kerosene became unavailable, and the Franklin stove became the sole provider of heat. However, the limited supplies of seasoned firewood in our community were snapped up within days after the start of the embargo. Unseasoned wood tripled in price and made a poor heat source, even when I supplemented it with deadwood that I laboriously gathered from the forests.

The gasoline crisis was just as severe, as it became close to impossible to buy any gas at all. I'll never forget the bitter cold morning I got up before 5 a.m., hours before dawn, to get in line at the gas station in order to ensure myself of a tankful of gas. But 52 cars had already gotten there before me, and many more joined behind. The temperature was minus 4 degrees F, with a fierce windchill factor. The cars formed a long line snaking far down the road, snorting exhaust fumes and wasting gas as the cold, stiff, hungry, worried occupants kept engines running in order to avoid hypothermia. Finally at 7 a.m. the gas station opened, and the cars began to inch toward the pumps. By 8 a.m. I got close enough to see a sign that read "3 gallon limit", but much worse was the sign that suddenly was placed two cars ahead of me: "No more gas today. Come back tomorrow".

This crisis of heating and transportation went on week after agonizing week. And yet, what happened on the Vineyard was minor compared to many other places across the United States where some people froze to death and food stores were emptied of vital supplies as people hoarded and no trucks came to replenish stocks. There were violent protests, strikes and blockades.

And then the whole mess was repeated in 1979.

The oil embargoes of the 70s were indeed shocking and frightening, profoundly threatening our sense of security. They provided a vivid demonstration of the danger of being dependent on foreign resources. All across the United States people were painfully affected as severe shortages and spiraling costs of oil and gasoline devastated their lives and economies at all levels.

In addition to the oil embargoes there were many environmental disasters. The oil tanker Torrey Canyon went aground in the English Channel, spilling millions of gallons of oil that ruined the ecology and economy for hundreds of miles of British and French coastlines. Another tanker went aground in Nantucket Sound right close to Martha's Vineyard, but by an incredible stroke of luck the wind blew the oil out to sea instead of onto our beaches and estuaries. There were oil well blowouts off the coast of Santa Barbara, in the North Sea, and the Gulf of Mexico. Later, 11 million gallons of crude oil spilled from the grounded Exxon Valdez, and it is reported that tenfold that amount is spilled annually in many smaller oil spills around the world.

Yet, we now use more oil than ever and are more dependent than ever on foreign oil. In fact, over 50 percent of U.S. oil consumption is imported, about half of that from the Persian Gulf region. The Gulf War of 1991 showed us the extreme danger, destruction, expense and suffering that is associated with that dependence.

Terrifying evidence, from Three Mile Island, Chernobyl, the Hanford Nuclear Waste Facility, and many hundreds of other actual or near-catastrophic incidents around the world, demonstrates the dangers of nuclear power, weapons and wastes.

There were increasing findings of polluted groundwater, ponds and lakes. Evidence showed that the major culprits were nitrogen and phosphorus, and although the blame was placed on laundry detergents, shorebirds and agricultural and landscaping fertilizers, there was mounting evidence that the major cause was actually human waste flushed through conventional septic and sewage systems.

Although some areas such as Lake Erie and the Charles River have indeed greatly improved, in these past 20 years groundwater and surface water has in many places continued to deteriorate. In Iowa, for instance, where most of the water pollution is caused by agricultural pesticides and fertilizers, over 40 percent of the groundwater is no longer safe to drink. When I visited Fairfield, Iowa, a few years ago, I was warned not to drink the tap water. I thought it would be safe to take a bath in it, but the water smelled so repulsive that I reluctantly took a quick shower instead.

I began to fear that it was perhaps inevitable that we would face unimaginable horrors within the foreseeable future and that humanity would foolishly drive itself and most of Earth's other wondrous creatures into extinction, either through freezing, starvation, and rampant disease, or with a global nuclear war caused by competition for our planet's land, water, minerals, oil and other limited resources.

But while I became increasingly aware and pessimistic about the problems, I was also finding out about possible solutions. Among the most important discoveries for me was the New Alchemy Institute on Cape Cod. Here a group of down-to-earth scientists had started in the early 70s to create a showcase of sustainable recycling design. They were making compost out of kitchen and garden wastes, growing worms in the compost, feeding the worms to fish grown in tanks, and then feeding the fish to people. They were using the "dirty" fish water for irrigating the gardens, which resulted in magnificent vegetables, and finally they completed the cycle by putting wastes from the garden and kitchen into the compost.

They were employing beneficial insects for controlling harmful insects and using windmills for pumping water and aerating the fisk tanks. They were using water and rocks as mass for storing the heat of the sun in greenhouses, and keeping flowers and vegetables, even bananas, alive through record-cold winters, without any backup heat.

I had long been nurturing my own compost and organic gardens and had read about experiments in recycling, nonpolluting, sustainable design around the world, but seeing it actualized so comprehensively and beautifully was believing it, and I yearned to adopt such methods. In the late 70s three major catalysts combined in my life to deepen my yearning and create the opportunity for transforming my dreams into reality.

Catalyst No. 1: You're-In-Charge

The first catalyst for major positive change in my life happened in 1977. By that time I had moved out of the house where my marriage broke apart and into a cozy little cottage in a clearing in the woods. This was a pristine paradise next to a marsh, among huge old hickory, beetlebung and oak trees, inhabited by birds, frogs, ferns and flowers, and filled with magic sounds and fragrances. Here I made the first of many startling discoveries that led me to believe that the situation for humanity and other life on Earth is anything but hopeless.

To some it may seem inappropriate and shocking to start out the list of these discoveries with a story about urine, but this is indeed how I first awoke to the fact that there may be ways of living sustainably. It truly surprised me when I discovered that this substance, so taboo and polluting in our society, is actually an excellent fertilizer. "You're-in-charge" began to put me in charge of my destiny.

This simple little cottage had no indoor plumbing and for the first year not even an outhouse. We managed, surprisingly happily actually, with boards placed securely over a hole dug into the ground at the base of a grove of tall elegant hickory trees. This was down a path a little distance from the cottage. Although we sometimes dreaded the trip out, we always came back happy from that outing no matter what the weather was like, rain, snow or shine. There was always some great Nature Show going on close by: a bird singing to us, or an ant trundling a stick four times its own size over the rough terrain of the forest floor, or sunshine sparkling on a perfect spider web draped with dewdrop diamonds. Anyone who has ever camped out in the wilderness can relate to this.

For peeing we usually stayed in the cottage, using a pot designated for this purpose, a technology ubiquitous in all sectors of society until so recently. Knowing that dog urine can kill grass and bushes, we disposed of our urine by first diluting it with water, about 1:10, then tossing it out here and there in order to avoid any damage and odors.

It never did smell, but I was in no way prepared for what followed when spring burst forth after that first winter. The first indication was from the wild lilies-of-the-valley that soon carpeted the forest floor. Here and there were patches of significantly healthier and larger plants. Then there was the little stunted bleeding-heart plant I had adopted and planted the previous fall. It quickly grew into an exceptionally large and splendid specimen which was amazing enough, but I really began to wonder what was going on when it then proceeded to bloom way past its normal stopping time in mid-July. In fact, it bloomed continuously up until frost took it in late fall.

Indian poke plants, normally no more than 5 feet tall, grew to over 10 feet tall, reaching half way up the second floor window. The spirea blossomed into the most extraordinary bridal bouquet display. These and other unusually vibrant patches of plants, such as St. Johnswort, Queen Anne's lace and black-eyed Susan, formed roughly a circle around the cottage, and I soon realized that these were the places where we had tossed the diluted urine. Could it be that this waste product was not just an ecological menace, but could actually be a beneficial fertilizer?

Most intriguing of all was the kinship and communication I sensed with these plants. The waste molecules from my body were being absorbed as nutrients by the living plants. I felt reincarnated while I was still alive. This liquid soon inspired new names: Peace-on-Earth... Urine-charge... You're-in-harmony... Aqua Vitae....

I started using this golden elixir purposefully, with astonishing results, as is evident in the photographs. As an experiment, I sowed some lettuce seeds directly into a growing medium of plain peat moss (practically void of any nutrients) and provided only diluted urine and a sprinkling of wood ashes as nutrients. These grew into full-sized healthy (and delicious) lettuce heads.

I began to understand some of the language of the plants. To describe the physical signs and the process in a greatly simplified way: when green leaves turn toward yellow-green, it means the plant needs more nitrogen. I then apply diluted urine, and within a day or two it begins to respond by becoming greener and more lush. At the first sign of the green turning bluish, accompanied by a subtle shrinking, closing attitude which means "enough nitrogen for now, thanks", I withhold the diluted urine until the plant goes through the bright-green full-open stage and begins to return to yellow-green again. Depending on the speed with which the plant is growing, the cycle is about two to six weeks.

I learned the hard way not to overdo it, especially with potted plants. One actually died from overdose. But there were only a few failures among the many plants that thrived. A friend lamented over a small maidenhair fern which had not grown one new leaf since she had bought it a year earlier. I suggested applying Peace-on-Earth, and after a month she reported that the plant now had five new leaves, more than twice the number it had before.

I went in search of information about the nutrient content of human urine, but could find nothing in the literature. Finally I found something, in a book that quickly got on my list of Most-Important-Reading: THE INTEGRAL URBAN HOUSE by Olkowski et al. of the Farrallones Institute in California, published by Sierra Club Books. Here a chart states that urine contains, by dry weight: 15 to 19 percent nitrogen (far higher than any animal manure), 2.5 to 5 percent phosphorus, 3 to 4.5 percent potassium, and 4.5 to 6 percent calcium. No wonder the plants were responding so happily.

The output of nitrogen per average person is about 10 pounds a year, over 6 pounds in the urine and another 4 pounds in the feces. As stated earlier, conventional septic systems release most of this nitrogen into the ground water, and this nitrogen then seeps unabated with the groundwater toward the nearest down-gradient body of water, whether 100 feet or five miles away. There it causes massive algae infestations which have a disastrous effect on the ecosystem of ponds, lakes, rivers and harbors.

Urine is usually sterile (unlike fecal matter, which always contains many kinds of harmful pathogens, such as staphylococcus and streptococcus, even from a healthy person), and therefore it appears that your own urine can be safely used for growing your own food. For a couple of years, before starting to produce vegetables to sell, I used no fertilizer other than diluted urine and ashes from the woodstove (using wood and waste paper as fuel) with great success, as the photographs demonstrate. Ashes, on the alkaline side of the pH scale, balance out the pH of the urine, which needs to be slightly acidic in order to avoid urinary tract infections. The alkaline ashes also help precipitate out the salts in the urine, so that they do not accumulate to harmful levels in the soil. In addition, ashes provide extra potassium as a plant nutrient.

However, I realized after some time that using just urine and ashes was not providing the soil with the "fodder" contained in compost, which is needed to keep up a high humus content, and this would eventually lead to a downturn in garden productivity. Compost, made from a mix of manure, food wastes, leaves and clippings, shredded paper, sawdust and other organic "roughage", enriches the soil with humus, so essential for increasing the water-holding capacity and microscopic surface areas that enable the earthworms and other visible and microscopic creatures to flourish. These creatures aerate and mix the soil, and their life and death cycles break down the minerals in the soil and thus make them available as nutrients for the plants. In other words, urine acts more like a chemical fertilizer, feeding not the soil, but the plants directly. So I decided to put the urine in the compost bin instead, which speeded up the decomposition process, and then to use this nitrogen-enriched compost to feed both the soil and the plants .

Important Note: I want to clearly emphasize here that for the Solviva Salad or any other food produced in the Solviva greenhouse and gardens I did NOT use urine or even the compost containing human urine. Why? Because Western society, especially the United States, has such an extraordinary taboo and prejudice against human body waste. This taboo is for a good reason, because if human waste is not managed properly, as was the case in the Middle Ages and beyond in Europe and still is the case in many parts of the world, it can contaminate water and food with pathogens that can cause disease. Smallpox and cholera epidemics killed millions. Underground sewer pipes and the flush toilet (invented in the mid-1800s in England) greatly reduced health problems stemming from bacteria and parasites present in human feces. But, as we now know, the methods that evolved in western culture for dealing with the wastewater proved to have disastrous effects on the environment and the economy, as I explain again and again in this book.

Catalyst No. 2: Meditation

While living in elegant simplicity and close communication with the plants in the forest, I became increasingly convinced that we can learn to live in harmony with life on Earth. But at the same time I grew ever more despairing over the destruction being perpetrated by modern man. What could I do to make a difference? How could I achieve the clarity of purpose that I observed in the birds, frogs, trees and flowers among whom I lived?

I began to learn about meditation and yearned for the stillness of mind that was said to be possible to achieve. A friend told me about the Insight Meditation Center in Barre, Massachusetts, and I signed up for a two-week retreat led by Joseph Goldstein. This beautiful center, previously a monastery, was filled to capacity with 125 seekers. Sitting meditation alternated with walking meditation, and these were interspersed with working meditation and eating meditation. The object was to be aware of every detail of every gesture in every second, for as many hours as you could stay awake, letting the thoughts go by without getting trapped in them, all in silence and with no eye contact. The schedule started before dawn and went into the night.

The first couple of days were agony. My back and legs were in excruciating pain, but this I could at least contol somewhat by changing positions. But I was horrified to be confronting my uncontrollable, incessantly grinding mind. Plans, ideas and insights were randomly mingled with nonsense and silly repetitive phrases, sabotaging my attempts to count my breaths or repeat a mantra in silence. Joseph reassured us that this was going on, more or less, in everyone's head, even his own, and I couldn't help laughing as I visualized the balloons of silent cacophony rising from the 125 heads in the Great Hall. We were urged to be very patient with our minds, just to notice the amazing scenes go by, without attachment or rejection, and to keep breathing, keep relaxing.

Gradually I started to experience silence in the mind and complete body relaxation, first for just brief moments, then for longer and longer stretches. Each silence was like a golden glow which, paradoxically, both filled and emptied my mind and suffused my body. On the second to last day of the retreat I experienced something I had never imagined could be possible. I was in that golden state of mind and body when I suddenly felt my body bounderies melt away, first my head, then down to the ends of my fingers and toes. I had no sensation of the size or whereabouts of any part of my body. My mind and soul were free, floating somewhere 2 feet or 200 million miles above. I was utterly free of effort, filled with bliss, barely musing that I could remain like that forever - and then what?... In this state I became aware of a quiet, powerful, brilliant presence, not a defined entity but rather a force imbuing everything everywhere, including the limitless me. And in this state I, having no previous experience with religion, silently prayed from the depth of my soul: God, what/whoever you are! Just tell me what to do! I'll do anything to help restore and protect this Garden of Eden....

The answer came promptly, slow and clear, from behind and above my left shoulder:
Do No Harm.

Suddenly I was back in my body and in the room. Tears were streaming down my face and my shirt. I had never before felt so completely happy and content. Do No Harm! How simple. Just live in ways that do no harm, live in harmony with life on Earth, and it will again be the Garden of Eden. Now I knew what to do with the rest of my life.

This event fit so well with my discovery of how urine could nourish plants and how this opened up sacred channels of communication. Soon after I returned home, I was talking not only with the plants, but also with the frogs and the birds who lived in the woods. They started to come into the house, where we shared eye/soul contact and conversations. The most extraordinary of these events was with a hummingbird. I had just finished meditating out on the deck and was in that golden state of mind, when I saw the little hummingbird close by, droning along from flower to flower, sipping nectar. I started talking to it in the rather silly, high-pitched, gentle singsong way that I had found effective for communicating with animals.

To my surprise and delight, the tiny creature sat down on a branch just a few feet away, looking at me. I flowed imperceptibly toward it, my hand outstretched, all along talking softly. Soon I found myself ever so gently stroking the little bird, who reacted by visibly relaxing, eyes half closed. Occasionally a dog barked in the distance, and each time the bird bristled and popped open its eyes, but a second later the eyes were again half closed, the body and feathers relaxed, leaning onto my finger. This went on for many minutes... could have gone on forever... and then what? There is indeed SOMETHING going on, beyond what we think of as "normal reality".

These kinds of experiences strengthened my commitment to living in ways that Do No Harm. But I needed more sunlight than was available at the site of the little cottage in the woods, so I decided to move back into the bigger house which I had left two years earlier. Although poorly built and drafty, it had great solar exposure. What I had learned about solar power intrigued me, and I wanted to incorporate it into my new way of living. Thus, with meager funds I immediately set about to increase the solar heating potential of the house.

The first thing I did was to knock out some of the south-facing walls and replace them with large windows. Next I built a deep indoor planting bed below this new long window wall and filled it with good soil from the summer garden. I transplanted dill, parsley, thyme, chives and marigold from the garden and sowed lettuce, spinach, chard, carrots, beets and radishes.

Then I laid up several cords of good seasoned oak firewood and installed the latest state-of-the-art airtight woodstove, especially paying attention to fire safety in every detail.

Then, as winter descended and the sun was pouring in through the windows, I sat back and watched the garden grow. And it grew ... and grew. The dill, for instance, grew to over 5 feet tall. I had always had houseplants, but growing food plants inside in the winter cold was a special thrill.

On a frigid, windy, sunny day in mid-winter my home was one of several open for a solar home tour. Busloads of visitors streamed through, oohing and aahing about the indoor garden and the cozy warmth, especially when they realized that the woodstove was cold and that all the warmth was provided by the sun.

The various pieces of evidence that had unfolded for me over the period of a year or so were powerful indeed. To me they were proof of extraordinary possibilities that I could not ignore. Although I continued my longstanding profession as a weaver, I became restless. My hands continued weaving the wool yarn into soft blankets and shawls, but my mind was weaving a different cloth. In the warp of the impending worldwide doom and gloom I laid in a weft of glorious opportunities for sustainable living.

I decided to open a restaurant to provide wholesome, organic food and also information about sustainable ways to live. I rented a defunct greasy-spoon joint in the center of Vineyard Haven, at cost and in exchange for cleaning it up. A bunch of friends helped remove most of the old grease, and I adorned the naked walls with houseplants, paintings and photographs, weavings and quilts that I brought from home. I offered Good Food for Body, Mind and Spirit, and called it The Rising Sun. I served one entree each night, salads and soups, fresh-baked Swedish bread, and killer desserts. I also offered the space as a forum for music and poetry, slide shows and movies. In the bleak winter on the Vineyard this place became an instant success.

That February we were experiencing a prolonged and severe cold spell, accompanied, as such cold usually is, by clear sunny days. I spent the mornings at home, doing paperwork by the hot woodstove. When it was time to leave for the restaurant, the fire had burned down to glowing embers. If I had stayed home, I would have just let the fire go out and the house would have been toasty all day with just solar power. Then, as the house rapidly lost its warmth in the late afternoon (it was drafty and poorly insulated, had no solar heat storage, and I had not yet made the movable window insulation), I would have relighted the fire for the evening and night. But on the days that I worked in the restaurant I was gone from 11 a.m. to 11 p.m., and, even though it was solar-heated during the day, the house and the woodstove were ice-cold when I got home. It took what seemed forever to build a fire and begin to warm up the stove. This was not very cozy.

But I learned what I thought at the time was a great trick: if I laid firewood on top of the morning's glowing coals, and then closed down the stove completely - the flue damper, the primary draft control in front, as well as the secondary draft control in the back - then the combustion of the wood was greatly reduced because there was no inflow of oxygen. In fact, instead of burning up, the firewood turned to charcoal. When I then returned home at night, I simply opened up the draft a bit, and as oxygen hit the charcoal and the remaining low-glowing embers, voila! there was an instant glorious fire. But what I thought was a great management technique soon proved to be disastrous.

Catalyst No. 3: Fire

On February 20 I was down at the restaurant as usual, baking bread and making soup. Suddenly a friend rushed in and reported that he just heard on the emergency radio in a store that my house was on fire. NO, that's impossible, they must be mistaken, it must be one of the neighbors', not mine, no way, my woodstove and installation must be the safest on the island, I left the house just a couple of hours ago and all was well, it can't be, not mine! As we raced in his truck the four miles to my home, I saw over the hill a giant column of smoke rising straight up into the sky. NO, NO, it cannot be my home!!

But as we neared the end of the driveway, I saw the unthinkable. No amount of writhing or screaming could erase the horrible truth: my home was totally engulfed in an inferno. Amid the crashing explosions and deafening roar I prayed not to lose the two items that seemed most precious: the family photographs and my daughters' early childhood drawings. It seemed that if they were destroyed, the footprints of our lives would be erased.

A life raft of friends took me to the house next door, and one of them told me to drink a whole glass of - whiskey! Indeed it did the trick of releasing me from the grip of hysteria. I finally accepted the reality that my home and perhaps all my belongings had been destroyed. As soon as the fire was out, we went back up to the site. Amid the stinging smoke, stench and grotesquely disfigured remains, I went straight to seek what I had prayed for. There, in the living room, protected under all the debris that had collapsed on top of them, were the family photo albums and in the back bedroom was the box of drawings, all miraculously protected. Though charred around the edges and smoked and soaking wet, they were among the very few things that were saved. No life had been lost: my daughters were away, my dog had been with me, the cat had been outside. Even a little mouse, who had lived in a large terrarium, had been seen by the firemen escaping along the charred, smoking rafters.

A deep joy and gratitude surfaced through the despair, increasing as I began to realize the opportunity that could be born out of this crisis: I could rebuild, this time starting from scratch instead of picking away at a drafty house. This hope was reenforced by a long-lost friend who happened to call a few days after the fire. He had lost both his home and his studio in a fire a few years earlier (fortunately my studio was in another location). But out of this catastrophe his life was reborn far better than it had ever been before.

But there remained a large question: what had caused the fire? It was obvious that the root cause was the fire in the woodstove, but not even the firemen could figure out how both the woodstove and the chimney had been ripped asunder by a massive explosion.

It was two months after this disaster that I learned of other houses in New England burned down in similar ways, previously unheard of. It was reported that the cause was gas explosions in these new state-of-the-art, super-airtight woodstoves. The explanation was that these new woodstoves, such as I had, could be closed down so airtight that virtually no venting could take place. When firewood is piled on top of a bed of hot coals and the stove is closed down really airtight, the wood heats up and explosive gases continue to emit from the hot wood, and having no oxygen, the gases cannot burn but instead accumulate to critical mass, like a propane gas tank. It was reported that a little back-puff of air down the chimney, plus a bit of glowing coal, is all it takes to ignite the gas and cause a massive explosion. Now I finally knew what had caused the fire.

Phoenix Rising

These three catalysts, the urine, the meditation and the fire, took place during the 18 months around my 40th birthday. A few months before, I had put a clipping on my refrigerator that read: LIFE BEGINS AT 40. I had certainly known happiness and fulfillment in the first 40 years of my life, but after these three events a very clear vision developed of what my post-childrearing life and mission would be about.

Soon after the fire I set about to design the ideal No Harm Home. I decided to reexamine all aspects of how we live and their effect on economy, security, health, environment and resources near and far. My wish was to build a house that greatly reduces the negative impact that is caused by the normal ways we heat and cool our homes, get our food, and manage our wastewater and solid wastes. I spent the next 16 months searching for the best sustainable and renewable designs and methods. I read everything I could lay my hands on from all over the world and talked with anyone who might have some knowledge. In addition, because I wanted to see for myself what was effective and what was not, I traveled to some innovative places around New England and California, and also to Great Britain where interesting experiments were going on.

HEATING

❦ ❦ ❦

I wanted to heat my home in ways that eliminate or minimize consumption of oil, gas or nuclear power. For one, I did not want to pay $1,200 to $2,500 a year for heating. For another, in my small way I wanted to demonstrate that there is no need for the United States to consume the energy equivalent of 150 billion gallons of oil a year for heating (25 percent of the total U.S. energy consumption).

I wanted to heat with solar power as much as possible, but I also knew I would need some backup heat source. I had vowed never again to have an airtight woodstove, not only because it had caused my home to burn down, but also because I felt suspicious about the smoke that comes out of the chimney from an airtight woodstove. It smells as if it releases far more toxic substances than the smoke from a hotter fire.

But having many millennia of genetic memory of fire as the only means of avoiding certain death during long, dark, cold Scandinavian winters, I still believed in a woodstove. So I decided to get a large Franklin stove, not airtight, but with doors that can close and thus greatly reduce the severe heat loss that occurs with an open fireplace. To minimize the risk of fire, and to capture and store as much heat as possible, I planned to set the stove in a massive brick and concrete alcove, with a thick concrete slab a foot above the top of the stove. I also planned to lay a coil of copper pipe on top of the stove and connect it in a revolving loop with the hot-water preheating tank. Thus, every time the fire was lighted, water would be heated for free.

During the intense study and design period after the fire, I made a discovery that was to contribute greatly to my growing understanding of solar design principles. At a gathering at New Alchemy Institute, on Cape Cod, I received some reprints from George Mokray, a fellow visitor and founding member of the Boston Solar Energy Association. These reprints contained information about a Professor Edward Morse of Marblehead, Massachusetts, who way back in 1881 made a remarkable discovery: by employing two basic laws of physics - 1) that the sun shining on a dark surface creates heat, and 2) that hot air rises - he created solar wall heaters, or solar thermosiphon panels, on the outside of south-facing walls, thereby successfully heating the rooms inside.

The Morse solar wall heaters consisted of an outer surface of clear glass and an inner black surface, with air space in between, mounted together in a frame attached to an exterior south wall. Heat was generated when the solar energy struck the black surface, and it was trapped in the box by the glass glazing. Vent openings top and bottom led from the solar box and into the house. By

The speaker had next experimented at his own house, where he had made a heater on a different plan. The corrugated iron was replaced by slate, and the glass was placed vertically outside it, just as in an ordinary window sash. The space between the glass and the slate was three-quarters of an inch, and the flue, or space behind the slate, in which the air ascended, was four inches deep and three feet wide. The room was about nine feet high, and measured about twenty-one feet by thirteen feet, its cubic contents being about 2500 cubic feet. At the foot of the flue was a damper which could be opened or closed from within. When open, the outside air could pass up through the flue and in at the top of the room; when closed, no outside air could enter, but the air from the lower part of the room could pass out into the flue and ascend entering again at the top. With the glass arranged vertically, it was found that in summer the sun gets so high that almost all the rays are reflected, and the slates are but slightly warmed. Fig. 5 illustrates the construction of this apparatus.

Fig. 5.

A A, Sun-heater. B, Glass. C, Slate. D D, Valves, by the operation of which the outer air may be drawn into the room above, or the air of the room may be drawn out below.

The following table shows the results of some observations during part of one day:—

Time.	Temperature at floor.	Temperature half way up to ceiling.	Temperature at mouth of heater.
8.45 A.M.	62° F.	64° F.	75° F.
9.00	62.5	65.0	80.0
9.30	63.0	65.0	83.0
10.00	64.0	66.0	85.0
10.30	64.5	66.3	86.0
11.00	65.0	67.0	92.0
12.00	67.0	68.0	90.0
1.00 P.M.	67.0	69.0	89.0
1.30	67.0	70.0	86.0
2.00	68.0	71.0	89.0

The following table, for another day, gives, in addition, the number of cubic feet of air discharged into the room, as calculated from anemometer observations:—

Time.	Temperature of room at floor.	Temperature at mouth of heater.	Cubic feet of air discharged into room.
9.45 A.M.	62° F.	91° F	3226
10.45	64.0	91.0	3326
11.45	66.0	95.0	4119
12.45 P.M.	66.0	95.0	3062
1.55	67.0	91.0	1299
2.45	68.0	80.0	

It appears from this table that between the hours of 9.45 and 1.55 there were 13,733 cubic feet of air discharged into the room, or over five times the cubic contents of the room, thus changing the air in about forty-five minutes.

Another experiment gave the following results:—

Time.	Temperature at floor.	Temperature at mouth of heater.	Cubic feet of air discharged into room.
9.45 A.M.	64° F.	87° F.	2777
10.45	66.0	97.0	2777
11.45	67.0	97.0	3870
12.50 P.M.	68.0	94.0	2928
1.50	69.0	92.0	

In this case the total quantity of air discharged into the room was 12,353 cubic feet in about four hours, or about fifty cubic feet per minute, the air of the room being changed once in every fifty minutes. This, moreover, was on a hazy day.

This method of heating can, of course, only be auxiliary to other methods, as it will be inoperative on days when the sun is obscured. When in operation, however, it has the advantages of freedom from dust, cinders, danger of fires, etc. The observations of the signal office in Boston show that in the five months from Oct. 1, 1884, to March 1, 1885, there were fifty-five clear and seventy-seven fair days, or one hundred and twenty-five to one hundred and thirty days during which the method could be used.

The most extensive application of the method has been to one of the rooms at the Boston Athenæum, where it has been in use several years. The box is forty-two feet long and six and a half feet wide; the area of heating surface being two hundred and seventy-three square feet. The room to which it leads is unfavorable for the use of the method, being covered above by a skylight, which rapidly cools the air within. Experiment has shown that, with a temperature of 55° in the sun at the mouth of the heater, the temperature at the upper entrance to the room was 97°, and that in four hours 312,984 cubic feet of air were discharged into the room, or 78,246 cubic feet per hour. Again, with a temperature of 31° in the shade, that at the mouth of the heater was 38°, and at the entrance to the room 76°.

The method which has been described will, it is thought, prove of use especially in the case of churches, halls, etc., which are only used occasionally, and by its action in ventilating will remove the damp and drowsy atmosphere often found in such places. The air will be good and fresh, and the apparatus will cost nothing for repairs.

In the discussion following Prof. Morse's paper, Mr. EDWARD ATKINSON said that, if by this means it would be possible to remove the bad air which collects in print-works and bleacheries, it would be of great benefit.

Prof. MORSE, in reply to a question as to the cost of the apparatus, said that the Athenæum heater cost $275 or $300, but could be built for much less now. The heater on his own house in Salem cost about $18. Regarding the amount of coal which the use of the apparatus would save, it had been estimated that to do the work accomplished by the Athenæum heater would require between twenty-five and fifty pounds of coal per day.

170 DAYS/YEAR REQUIRING HEATING
x 25-50 POUNDS COAL PER DAY
4250 - 8500 POUNDS COAL PER YEAR.

opening these vents, the solar heat created an updraft. Morse reported a temperature rise of 42 degrees F between the air entering the box from the room through the bottom vent and the air entering the room at the top vent.

This information seemed to me to hold the Golden Key to effective solar design, offering the solution to a major portion of our energy problems. Morse had invented or discovered this design in 1881, and the big question in my mind was: why was not every south-facing wall and roof built like this?

Perhaps the reason this beautiful design did not fly was that around the same time, oil was gushing out of the ground and became available as a cheap heating fuel. I can well imagine Morse's despair as he foresaw at least some of the devastation that would follow if society's entire infrastructure became dependent on drilling, burning and competing for Earth's finite supplies of fossil fuels.

This was the most elegantly simple solar devise I had seen yet. Other solar designs had serious drawbacks and limitations. Some called for roofs with wide overhangs to prevent summer overheating. But such overhangs seriously reduce the possibility of growing food plants indoors, because most require full light from above as well as from at least one side. Overhanging roofs also limit architectural style. Other solar designs I saw consisted of unattractive and expensive collectors attached on top of roofs, vulnerable to damage by wind and snow. In general, solar design was getting, and still has, a bad reputation as being unreliable, inconvenient, unattractive, too hot in the summer, too cold in the winter and (in spite of tax credits) very expensive, as well as resulting in leaky roofs, bleached out furniturea and lukewarm showers.

It seemed clear to me that by adapting and applying the Morse design to any building with solar exposure on wall or roof, whether a home, high-rise apartment building, school, or corporate building, most of the needs for heating air and water could be provided by the sun. If this was possible, then there was hope indeed.

Soon I set about to create my own experimental version of Morse's solar heater, to prove it to myself. I built it inexpensively on a wall of the barn I was temporarily living in, with black tar paper for collection surface and clear polyethylene for cover. I substituted Morse's straight, vertical air-flow path with a winding path constructed with horizontal 2-by-4's painted black, thinking this might create even more heat because the air had to travel much further between the inlet and the outlet. I completed the heater by sealing the plastic cover around the edges at about 11 a.m. on a sunny cold day in late March. I read the thermometer at the bottom vent where unheated, uninsulated air entered the solar heater: 32 degrees F. By the time I got upstairs, some four minutes later, there was a stench of hot tar rushing into the room with the solar-heated air, and the thermometer at the top of the collector was already reading 110 degrees F, a 78-degree temperature rise. Fearing another fire, I grabbed the fan I had waiting close by and stuck it into the top vent to pull the air faster through the solar heater, and soon the temperature of the incoming air was down to a safer 90 degrees F, still a 58-degree rise.

Although this was a very crude application of the Morse heater, it clearly proved the design to be at least as effective as Morse had described. In fact, it seemed that substituting his straight air path with a winding air path had increased the solar heat gain.

WARNING: Do not build a solar heater with such materials. It is definitely a fire hazard!

Amazingly, no one to whom I showed this design back in '79 had ever heard of anything like it. No architects, engineers, solar designers, manufacturers, or builders were incorporating any such simple, logical principles. Since then, I have seen minor applications of solar thermosiphon panels, but the way others have adapted the technology evidently is not efficient.

My developing house plans soon included Morse's invention. In the drawings the whole roof became a Morse heater, with fans to convey the solar-heated air down into heat-storing mass in the foundation. The mass I planned would consist of rocks and water, as well as the block foundation walls, well-insulated on the outside. The air would then return through vents leading to the north rooms and also into the greenhouse along the south wall and then, through a filter, back into the solar roof, thus creating a continuous loop of sun-warmed air circulation. I also planned to incorporate EPDM rubber piping into the solar roof, for preheating hot water, which along with a coil of copper water pipe on top of the woodstove would heat water in a preheating tank. From there the preheated water would go into a regular electric water-heating tank, which would ensure a dependable supply of hot water at all times, not only when the sun is shining or the fire burning. I, like most people, want dependable comfort.

I planned for the greenhouse/sunroom to be fully integrated with the living space, without the separating walls and doors that I saw in other solar designs. In order to prevent heat loss, I planned to make movable window insulation. I figured that the indoor garden would be far healthier and more productive if it was kept close to ordinary human comfort levels.

I took my drawings to be reviewed by several architects, solar designers and engineers, and they all stated various reasons why such a design could not work: the house would get too cold; it would get too hot; mildew and bugs would fill the house. Worst of all, one of them said, and as I had already seen in my crudely made Morse solar wall heater, it could constitute a fire hazard. I got further negative feedback when I applied for a federal grant which was being offered for innovative energy-efficient building design. The grant was denied, and indeed the design was given a poor rating by the examiner, who clearly did not understand the principles of my designs.

Even though I knew that many of the design components for my new home were radically different from accepted norms, I had it from strong inner and higher authority that they would work, and I stuck to my faith. (The fire plus daily meditation had greatly strengthened my intuition and powers of concentration and insight.) However, I did make a compromise: I reduced the solar portion of the roof to just one third of the total south roof area. In retrospect I wish I had stuck to my original plan of making the whole south roof solar, for it turned out to work splendidly. I of course took extraordinary precautions to prevent overheating of any flammable components.

My design called for two fans which would circulate the solar-heated air through insulated ducts leading from the solar roof and down into storage in the crawl space. I consulted with a heating/cooling engineer to help determine what size fans I would need. He measured the solar roof and the ducts, made some calculations, and declared that the system would require two huge expensive fans and that the design was worthless because it would use up more energy for electricity than the energy provided by the sun.

My intuition told me that so much power would not be needed. I happened to have two ordinary $20, 20-inch, 75-watt room fans. I installed them into the ducts and was then delighted to see that on sunny days, even extremely cold days, these simple, low-cost, low-watt fans pulled air heated to over 100 degrees from the solar roof (with 12-foot long, straight-run channels) and into the storage mass in the crawl space. If I had followed the advice of the professional engineer, the air in the roof collector would have been sucked through so fast that it would not have had time to be warmed up by the sun, and it would have cost a great deal more money. Now, 17 years later, the same fans are still faithfully doing their work, going on and off in response to the same thermostat.

Here it is appropriate to pass on a lesson which I have learned again and again over the years: be careful about following advice given by "experts". You need to plug in your own mind, intuition and common sense, rather than indiscriminately swallowing judgment rendered by an architect or engineer, especially from those who call themselves solar or environmental design experts. The entire methodology of standard ways of calculating solar design, or any engineering design, is seriously flawed, because there is more to it than just following established tables and charts in books and computers. The designs must be "felt" rather than just depending on calculations and formulas. If what the experts say does not make sense to you, and they cannot satisfactorily explain it to you, get other opinions until you find someone who does make sense to you. In my case, I could find no one who understood. I was on my own and just had to trust my intuition and experience.

Sixteen months after my old home was destroyed by fire, the new one started rising, and nine months after that, I moved into my unfinished home. The water-heating pipes that were placed in four of the 16 sections of the solar roof circulated solar-heated water to the preheating tank, and it soon became apparent that they reduced my water-heating bill by some 80 percent.

❧ ❧ ❧

COOLING

I had planned all along to have some form of summer shading to prevent the greenhouse and the rest of the living space from overheating. For the long term this would consist of a grape arbor to be trimmed back each winter, but for this first summer I would simply roll out bamboo screening over the glass roof of the greenhouse. However, I kept delaying doing that job because, to my surprise, even when summer heat waves descended and the sun was pouring into the greenhouse, the house was not overheating. I had made yet another major discovery about the solar roof design.

Even on hot sunny days my house remained comfortable, not overheating even in the fully exposed sunroom. The huge broccoli plants thriving in full sun in the indoor garden that first summer were silent but eloquent witnesses to the fact that this solar-heating roof was also providing highly effective solar cooling.

Since I did not want to fan solar heat into the basement in the summer, and since I knew that the solar roof would dangerously overheat without air circulation, I opened up the large vents of the plenum at the top of the solar roof. The hotter the sun shone on the collector roof, the faster the solar-heated air rushed out through the vents, consequently pulling the house air out with it. Replacement air then came in through the open windows all around the house. Thus, to my surprise, I came to discover that not only was this solar roof a very effective air and water heater, but it was also an excellent summer cooling device for the whole house. Consequently I never did plant the grape arbor I had planned, nor did I roll the bamboo screening over the greenhouse glass roof.

The "solar chimney" concept for solar cooling was experimented with in the 70s, but all accounts that I read or heard about were pronounced to be failures, incapable of providing adequate cooling for a greenhouse. As the name implies, these were structures built as a chimney, a tall separate stack about 4 feet square, with glass or clear plastic on the south side and the interior north, east and west walls painted black. Because they were tall, they were a menace, vulnerable to collapse in severe storms. They were also expensive.

I can see why my roof design works infinitely better. It has a far larger and better exposed solar collection surface and thus generates more heat, which in turn results in more updraft to pull out more of the air from the rooms below. Also, my solar-cooling device cost nothing to build, as it is simply a free fringe benefit, part of the solar roof. In addition, it has not been affected by storms: over the last 17 years it has survived numerous violent storms and hurricanes without a scratch or rip.

The most important implication of this solar roof is that you can have a greenhouse attached to your house, with full overhead clear glazing, without risking overheating. This makes possible a thriving, highly productive indoor food garden year-round, because the plants and the whole ecosystem receive light not only through the vertical glazing, but, equally important, also from above. Vertical glass and a few skylights in the roof are fine for growing plants that like partial shade, such as any of the rainforest plants, but most food plants are grown in full sun and therefore need full overhead glazing in order to achieve high and prolonged productivity.

Most other designs for attached greenhouses try to solve the overheating problem in one or more of the following ways: fans, which are noisy and expensive; overhanging roof to prevent summer sun from entering, which, as I explained, does not provide enough light for food production; exterior movable summer shade cloth, which is unattractive, a nuisance and subject to wind damage; or by what I had originally planned, an arbor. But an arbor requires conscientious trimming back and cleaning of the glass in order to adequately allow penetration of precious winter light.

This solar-heating/cooling design, coupled with the comprehensive subfloor heat-storage designs I have since developed, can be adapted to fit on or adjacent to almost any existing or new building in any location in any climate. We can thereby reduce the heating fuel requirements of any building by 80 percent or more.

❧ ❧ ❧

WASTEWATER MANAGEMENT

While designing my new home it became clear to me that I did not want a standard septic system. I had learned about the large amount of nitrogen that is contained in human waste, and that standard septic systems do not process this nitrogen but instead release most of it into the groundwater. And I knew that the nitrogen does not biodegrade or diminish in the groundwater, but instead travels unabated with the groundwater to the nearest lagoon, pond or harbor, no matter how far away. Once there, the first plants that react to the nitrogen enrichment are the algae, which consequently multiply explosively, thereby causing devastation to our ecosystems.

There are many telltale signs of nitrogen pollution: thick masses of billowing algae "smog" covering the bottoms of ponds, choking out desirable aquatic flora and fauna, and killing the shellfish beds; slippery, smelly beaches covered with slimy green algae at low tide; the horrible stench that periodically envelops towns such as Vineyard Haven, caused by masses of algae putrifying on the beaches around the harbor; flows of living green and dead brown algae covering vast surface areas of ponds; and large festering sores on beautiful tidal marshes, choked by thick layers of dead algae. For many years these algae problems were blamed on geese, ducks and swans, and on farms and lawn fertilizers, but in fact these sources pale in comparison with the contribution from septic systems.

Standard on-site septic systems are constructed in accordance with the current laws of man, but these laws are in serious violation of the laws of Nature. In Nature all waste products - leaves, manure, dead animals - fall to Earth's surface. They are then rapidly absorbed and processed by the different decomposers who live in the top few inches of the skin of our planet, including earthworms and pill bugs and myriads of microscopic organisms. I am not advocating that we dispose of our wastewater on the surface, because this exposes pathogens which could lead to disease. But we can, and must, dispose of the wastewater within the top 12 inches of the surface, thus making it possible for the nutrients to be processed by living plants. I knew I did not want to waste the precious nutrients contained in the wastewater. I instead wanted to use them to enrich the surrounding trees, shrubs and flowers.

Many people think that just changing to low-flush toilets solves the pollution problem, but it is obvious that this in no way reduces the amount of nitrogen injected into the groundwater: although low-flush toilets do save water, requiring only 1.6 gallons or less per flush instead of the standard 5 to 7 gallons, they still flush exactly the same amount of nitrogen into the same standard septic or sewage systems and thus cause just as much pollution. Back then I thought we simply had to do away with flush toilets and replace them with various kinds of waterless toilets.

I researched different composting toilets developed all around the world, but discovered they all had problems such as mechanical failure, flies, odors, great expense, or limited capacity. Some had wide chutes leading to a composting chamber far below, evoking terrible images of an unwatched toddler or puppy falling in, head first. I did not want to have to padlock my toilet. Some systems used so much electricity that it seemed that any benefits gained were outweighed by the increased need for oil, coal or nuclear-generated power. For instance, one composting toilet required about 2 kwh/day for fan and heater, which is about 700 kwh/year (comparable to a refrigerator), which at our current rate of 15 cents per kilowatt-hour is $105 per year. So I designed my own composting toilet as well as graywater management system, based upon what I had learned while living in the little cottage in the woods.

When I went to the local board to obtain my building permit, I was fully prepared to have to defend what I felt was my right to build and test an innovative toilet system that I believed would be better than standard septic systems from both public health and environmental standpoints. I knew that regulations are rigid in these matters, but I was prepared to "go to the Supreme Court" if I had to.

But to my surprise and delight I was given the building permit without any questions about septic plans. Apparently the permitting board figured I would hook into the preexisting septic system because the new house would be located adjacent to the one that had burned down. Thus I had the opportunity (for which I have always been grateful) to experiment and to discover that it is possible to have a toilet that uses no water, that is easy to maintain without odor or fly problems, and that costs very little to construct. I also discovered that it is possible to have a graywater management system that causes no pollution, costs no money, greatly enhances the landscaping, requires no maintenance, and has no problems whatsoever.

The Solviva Compost Toilet

I created two compostoilets, one downstairs and the other upstairs. The toilet consists of a polished mahogany cabinet with a comfortable seat and a tight lid. Below is a 20-gallon plastic barrel holding tank. Adjacent to the toilet is a hinged compartment containing a cover material that consists of sawdust or shredded leaves mixed with some compost. Thus, instead of flushing with 5 or 6 gallons of drinking water, a scoop of this cover material is all that is needed to immediately eliminate any odors.

The sawdust or shredded leaves in the cover mix perform the task of bulking and aeration. Also, their high carbon content bonds with the high nitrogen in human waste, resulting in excellent compost. The sawdust must be easily biodegradable, such as pine, hemlock, spruce, fir, maple or oak, not rot-resistant wood such as cedar, locust, teak, mahogany, or any wood treated with preservatives. The sawdust must also be fairly fine but not powder-fine. The consistency generated by a tablesaw is perfect. If leaves are used, they must be finely shredded. Old leaf mold works well. I do not recommend peat moss for compostoilets, because the fibers absorb so much moisture and swell up so that air circulation and oxygen penetration is greatly reduced.

The best, and also the most convenient, compost to mix with the cover material is the finished toilet compost, because it contains the organisms that have evolved to best handle this particular ecosystem. But to start with, any good compost will do fine.

The toilet barrel has a drain through which the urine flows into a pipe that leads to the outside composting chamber. The urine adds valuable nitrogen and moisture to the compost, both of which speed up the composting process.

When the toilet barrel is three-quarters full (this takes about four weeks for a single person), the whole barrel is removed from the toilet cabinet and taken to the compost chamber area. The easiest management system entails topping this barrel off with compost from the most active compost chamber, rich with earthworms and other decomposing organisms, and covering securely with a lid. After a short period of time, about two to three weeks in the summer and five to seven weeks in winter, the contents of the barrel will be fully composted and have the wonderful fragrance of good humus. After emptying the barrel into the compost bin, it is clean and ready to be used again. The changing process takes less than five minutes, and the different aspects of the whole system smell like good earth. Zero nitrogen enters the groundwater because the compost chamber is covered by a roof to prevent rains from leaching the nitrogen into the subsoil and groundwater. The nutrients are all absorbed and processed by the various carbon materials and organisms in the compost sponge, resulting in top-quality compost.

My two compostoilets and the compost chamber cost just a couple of hundred dollars to build. The system has been functioning perfectly for 17 years now, and there is no reason why it would not be functioning equally well for another 100 or 5,000 years. There is no leach field that can clog up, as in standard septic systems. There are no plumbing pipes to plug up, as with flush toilets. There is simply nothing that can go wrong, except of course by human carelessness. For instance, if you don't use the cover material, it may smell. If you do, it doesn't. If you let the drum get too full, it is heavy to move. This compostoilet can be left unused for months on end, or can be used continually by one or any number of people. For large numbers of people, the barrel simply needs to be changed more often, but, as I said, that takes less than five minutes.

To some people the concept of a waterless toilet is really threatening, especially to men. There have actually been times that men have asked for the bathroom, and when told that it is a composting toilet, they swiveled right around, and while avoiding eye contact, said: "That's alright. I can wait." Whereas most women go right in, stick their head in and say: "It smells wonderful and looks so beautiful and clean. How is this possible?!" Children love it. They just sit there and glow, and soon "need to go" again. Leave it to kids to have the right instincts. For many people, using these compostoilets has been a deeply transformative experience, especially for those who have had a chance to participate in the changing process. People are elated when they see human waste transformed so effortlessly into a truly lovable product.

Graywater Purification

Even with compostoilets, I still had wastewater to deal with, namely that portion of the wastewater referred to as graywater. This is all the wastewater except that which comes from toilets. Current regulations sometimes allow compostoilets, but generally require standard septic systems to handle the graywater, although the size of the leaching field or pit can be reduced by 40 percent compared to households with flush toilets. Even these reduced septic systems require extensive excavations, and usually cost around $6,000 to $8,000, and in many cases far more.

In my new home I did not want to put my household graywater into a standard septic system, for reasons of pollution, landscape destruction, high cost, as well as waste of good irrigation water. Instead, I designed what I call a "fully upgraded", "bio-benign" septic system for purifying and utilizing the graywater.

The graywater pipe runs from the house down a short slope and ends in a shallow depression filled with wood chips and leaves. Grasses and wildflowers and some sumac bushes, Indian poke and a cherry tree already grew around this area, and I added a Rosa rugosa bush, a dogwood, a Norway spruce, and a white pine. These are my sewage treatment plants, and it is a most beautiful area at any time of year. All the wastewater from the house (occupancy ranging from 1-10 people) has exited here in this same spot for the last 17 years, and there has never been any wastewater overflowing to the surface, no scum, slime or odors, and no flies or mosquitoes. The wastewater just seeps through the topsoil, and the roots of the surrounding trees and bushes, grasses and flowers know just where to go to find what they want. As the pictures eloquently testify, these plants are not only surviving, they are indeed exceptionally healthy and beautiful specimens.

Now, you might think that I have been using only special biodegradable soaps and avoided any toxic stuff. On the contrary. I felt it was important to use "normal" household substances, in order to know whether this greywater management system could function in the "real" world, where few people are willing to spend extra money for special "biodegradable" products.

I have calculated that over the last 17 years, the zone of influence of my graywater management system has received and successfully processed approximately the following: 45 gallons of chlorine bleach, 500 pounds of ordinary laundry detergents, 8 gallons of detergents for washing dishes by hand (I don't have a dishwasher, so I cannot vouch for such detergents), 15 gallons of shampoos (including dandruff shampoos which, I have been told, cause havoc in standard septic systems), 5 gallons of hair conditioners, and umpteen pounds of toothpaste. In addition, I have dyed some 3,000 pounds of wool in my weaving studio, using aniline dyes and approximately 50 gallons of vinegar. (Note to those who find it strange that I use aniline dyes instead of "natural" dyes: I switched to aniline dyes when I found out that the mordants - salts of chrome, copper, tin and aluminum - that are needed with most dye plants caused more damage to the environment, my plumbing and me than the aniline dyes and vinegar mordants.)

Last summer this system was submitted to the ultimate test: the bathroom sink clogged up for the first time, and, after asking the plants to please brace themselves, I poured down a whole bottle of standard drain cleaner, to test the system with "real world" stuff. To my relief, none of the plants exhibited the slightest sign of distress. But that's really pushing it. Next time I will just use boiling water which, I am told, works just as well to clear the drain.

A standard septic system is incapable of breaking down most of the undesirable substances contained in regular graywater, but instead releases them into the groundwater. The Solviva graywater system, on the other hand, greatly reduces contamination of the groundwater by neutralizing these substances in an environment containing a mutually beneficial combination of oxygen, sunlight, composting organisms, humus-rich topsoil and plants. The system is trouble-free and will most likely last "forever" without any maintenance other than standard landscape trimming.

The Solviva Composting Flush Toilet

Over the years I came to realize that very few people will adopt compostoilets any time soon, no matter how convenient, economical and clean they can be. This realization made me quite pessimistic about the potential for stopping the vast amounts of water pollution currently caused by standard septic systems. So I decided to design a method of combining flush toilets with composting. I am happy to report that I have discovered ways to keep the flush toilet AND to prevent pollution to drinking water and aquatic ecosystems.

In September 1995, with capable carpenter and plumber, the upstairs compostoilet in my home was removed, and a regular 1.6-gallon flush toilet was installed. The toilet drainpipe empties into an enclosed Compostfilter, a box, 3 feet tall and 4 feet wide, built of wood against the outside wall of the house. The top and the front of the box can open for servicing and inspection. It is lined with plastic to prevent rotting of the wood and the house shingles, and it is insulated with 2-inch rigid foamboard to prevent freezing.

The box is watertight with a drain on the bottom and is divided into two connected compartments. Each compartment was filled with Biocarbon mix consisting of the right type and proportions of partially composted leaves and wood chips. I also installed 3,000 earthworms which I ordered by mail from a worm farm in Georgia, for $19. I call this box the Compostfilter. The toilet waste pipe has an elbow at the end that empties into the box. This elbow can be flipped to empty into one or the other of the two compartments. The solids, including the toilet paper, are retained in the compost box, while the liquid quickly seeps through the Biocarbon mix, exits through the drain and into a sloped pipe that leads to a series of ground-level Greenfilters.

I started flushing into the first compartment as soon as the installation was complete, September 20. Throughout the first cold winter, even through subzero conditions, the temperatures in the box stayed above 55 degrees F, without any heating source other than the low-temperature (mesophilic) composting process itself. My plan was to flip the toilet waste pipe over to the second compartment when the first one was filled. However, to my utter surprise, it never did fill up, because the process of decomposition has reduced the volume faster than it has increased by the additions of the daily flushings. Seven family members were at my home for a week over that first Christmas, all using this toilet because I wanted to provide a real stress test. The system continued composting and draining reliably.

After my family left, I flipped the elbow of the toilet waste pipe to drain into the second compartment, in order to get an accurate reading of how long it would take for the solids to fully decompose. To my amazement, I found that within one week there was nothing recognizable left in the first compartment, nothing but rich earthworm castings and healthy earthworms, which by this time had multiplied to hundreds of thousands of all sizes. When warmer weather returned I found that full decomposition was achieved in less than four days.

The truly astonishing fact is that the more I put into the box, the less there is in the box. As I write this, in October 1997, 25 months have passed since the installation, with usage ranging from one to seven people. This box has by now received approximately 2,500 flushes, 140 rolls of toilet paper, and 35 cubic feet of leaves and woodchips (the original 12 cubic feet plus periodic additions). The action of the earthworms and other bioorganisms has reduced it all to about 9 cubic feet of magnificent earthworm castings. Even with all my previous experience I never would have expected this could be possible.

When in its normal closed state, this compost chamber is absolutely odor-free. When you open it, it smells mildly like a stable. It has never generated any flies, and no earthworms have ever ventured outside, probably because this box is clearly Earthworm Heaven.

The liquid seeps quickly through this Compostfilter and runs via drain and sloped pipe to a series of Greenfilters. The first is a shallow growing bed filled with the same Biocarbon mix plus sandy topsoil and healthy plants, lined with a waterproof membrane, with a drainpipe that leads downhill to a second Greenfilter, same as the first. From there the liquid drains into a 20-gallon pump chamber equipped with a float switch-controlled sump pump that periodically pumps 15 gallons of effluent into a perforated pipe installed in the third Greenfilter. This is a flower bed, and the pipe is installed in a layer of wood chips 6 inches below the surface. From there the water perks through the subsoil, where any remaining pathogens are destroyed.

Thus, all the wastes flushed down the toilet are transformed into excellent earthworm casting compost and irrigation water that benefit the landscaping. The nitrogen is absorbed by the Biocarbon filter materials and the plants and does not leach down to contaminate the groundwater. I have also conducted other experiments with different versions of my Biocarbon filter, one for treating effluent from a standard septic tank, the other for treating septage and sludge pumped from numerous different septic tanks, including restaurants, businesses and homes. In the case of the septic tank effluent, lab tests showed a 90 percent reduction of the total Kjeldahl nitrogen, from 86 ppm to 8.1 ppm, and a 96 percent reduction of the ammonia-nitrogen, from 77 ppm down to 2.5 ppm. In the case of the Biocarbon septage treatment filter, total nitrogen was reduced 88.2 percent, from 152.34 ppm down to 17.81, while BOD was reduced from 607 ppm to 59 ppm, and COD from 640 ppm to 85 ppm. In both cases the flow-through took less than 10 minutes, and the foul odor was totally removed.

These results are nothing short of astonishing, to sanitation professionals and lay people alike. The Solviva Biocarbon filter systems demonstrate that we can manage our wastewater in ways that cause 90 percent less pollution than systems currently required by the state, and, in many cases, at far less cost.

Because of these successful experiments, I applied for a permit from the Massachusetts Department of Environmental Protection (DEP), to install Solviva Biocarbon wastewater filters in the "real world". After eight months of agonizing permitting procedures, the DEP finally gave me permission to install 15 pilot projects in Massachusetts, allowing a daily flow up to 10,000 gallons for each system. The first system has just been completed and is up and running, at the Featherstone Meetinghouse Center for the Arts in Oak Bluffs, and testing will begin soon. The second system is in the permitting process, for the Black Dog Tavern, a busy restaurant right on the Vineyard Haven harbor, that produces 4,000 gallons of wastewater per day. We are hoping to soon get the required permits, for the simple reason that there is not even the slightest possibility that the Biocarbon filter system could cause more risk to environment or public health than the current septic system. This project will tackle the most serious wastewater problem in town, and, if proven successful there as it has in the smaller experiments, will demonstrate that business or residential on-site septic systems even in the densest areas can be retrofitted with Biocarbon filters. This will render unnecessary the central sewage treatment facility currently being planned, and as a result the town will save millions of dollars.

Based on my accumulated experience, I believe sanitary, trouble-free, nonpolluting wastewater management is achievable in urban, suburban or rural environments, whether in hot India or cold Alaska. The Solviva Biocarbon filter systems can save thousands for a family, millions for a community, billions for a nation, as well as prevent the use of toxic chemicals, compared to septic and sewage systems built in accordance with today's regulations. In addition, important fertilizer resources and water are saved and made available for beautifying the landscapes.

I have developed designs for city high-rise buildings, in which low-water toilets flush into Compostfilters set in 30-cubic-yard roll-off containers located in the basement. These Compostfilters retain and digest the solids, while the liquid is pumped through a pipe leading to Greenfilter flower beds in a nearby park. If no green area is available close by, the filtered effluent can flow into the city sewage pipes, greatly reducing the load on the central sewage treatment facility. The Compostfilter roll-off containers are removed periodically to one of several composting facilities, which also process all other compostable wastes produced by the city, such as food wastes, leaf and yard wastes, diapers, low-grade wastepaper, and shredded construction wastes (together comprising some 70 percent of the solid waste stream). Here the high heat of thermophilic composting processes, combined with solar heat, can render the wastes into safe, sanitary, organic compost fertilizer with a market value of millions of dollars per year (Big Apple Black Gold, Boston Black Gold...)

Temperatures and Time Required to Destroy Common Pathogens and Parasites:

Organism:	Death at:	C	F
Salmonella	60 minutes at	55	130
	20 minutes at	60	140
Shigella	60 minutes at	55	130
Escherichia coli	60 minutes at	55	130
Brucella	60 minutes at	60	140
	3minutes at	65	150
Micrococcus	10 minutes at	50	122
Streptococcus	10 minutes at	54	130
Mycobacterium tuberculosis	20 minutes at	66	150
Corinebacterium diphtheria	45 minutes at	55	130
Entamoeba histolytica cysts	10 seconds at	55	130
Taenia saginata	5 minutes at	55	130
Trichinella spiralis larvae	instantly at	60	140
Necator americanus	50 minutes at	45	110
Ascaris lumbricoides eggs	60 minutes at	50	122

Source: Golueke and McGauhey.

Composting Organisms:

Psychrophilic Organisms work at	28-60 F
Mesophilic Organisms work at	50-100 F
Thermophilic Organisms work at	100-170 F

A teaspoon of living earth contains some five million bacteria, twenty million fungi, one million protozoa, and two hundred thousand algae. No human can predict what vital miracles are locked in this dot of life, this stupendous reservoir of genetic materials that have evolved continuously since the dawn of life on Earth.

One pound of topsoil has as much surface area as the whole state of Connecticut.

From Septic Tank Practices by Peter Warshall

One gram of living compost can contain 10 billion bacteria belonging to several thousand species, almost all of which are still unknown to science. E.O. Wilson

Easy Composting.

֍ ֍ ֍

The plans for my No Harm Home included composting all of the household organic wastes, including the toilet wastes, and I wanted to do it in an attractive, safe, convenient way. So I built a sturdy composting system, consisting of three adjoining bins, each 4 x 4 feet square and 3 feet high, all covered over with a shed roof. I located this structure about 75 feet away from the house because I feared the possibility of flies, smells, and even (horrors!) spontaneous combustion from high composting temperatures.

Little did I know that the compost structure could have been placed adjacent to the front door, because it is such an attractive part of the landscape. It is absolutely odor-free. It does not generate flies, because it has a permanent population of fly parasites, which I initially installed and which has since been self-generating. There is certainly no danger of combustion, because it never even gets hot - the composting process is mesophilic (cool), rather than thermophilic (hot), because the contents are added a little at a time. And there are beautiful flowers and vegetables growing in it April through November, self-sown from the waste materials that have been dumped into the compost bins.

After the first 6 months of loading in various organic wastes from the kitchen, the indoor and outdoor gardens, as well as from the two compostoilets, the left-hand bin was full, and I started using the right-hand one. About a month later I went out to turn over the contents of the left-hand bin for the first time, into the middle bin. But I discovered, to my amazement, that the materials were already fully composted. According to books and my previous experience, a compost pile needs to be turned over a couple of times in order to break down all the material. But now, without a single turning, and faster than I had ever seen before, the materials tossed into this bin had been transformed into the most beautiful, fine-textured, homogeneous, friable compost with a wonderful earthy fragrance. There was no sign of any toilet stuff, or even orange or banana peels or eggshells. The only things recognizable were coffee filters and avocado pits. I sent samples to a soil test lab, and the results came back showing excellent balanced nutrient and PH values, and no raised levels of copper, lead, aluminum or other undesirable substances.

This compost makes a great soil amendment and a beautiful mulch for bushes, trees and flowers. I do not use it for growing food because it contains humanure and therefore has the potential for recycling human disease pathogens. However, as I will explain in the section on Compost Pasteurization, that danger can probably be eliminated.

֍ ֍ ֍

Compost Pasteurization

Human fecal waste, even from a healthy person, contains pathogens that can cause diseases, and this is good reason for treating it with great care. Standard septic systems do not actually kill these pathogens, but most of the bacterias are eliminated as the effluent percolates through just a couple of feet of the subsoil beneath the septic system. Central sewage treatment facilities accomplish this task with toxic chlorine, or with UV radiation.

Humanure, or nightsoil, has long provided soil fertility all over the world. I have seen pictures of country roads in China, dotted with outhouses elaborately decorated in the hopes of attracting passing travelers and their valued deposits, and thus gain extra nutrients to apply to the fields. China has some of the oldest and most productive gardens in the world, many having been in continuous uninterrupted food production for thousands of years, since the dawn of agriculture. However, the use of improperly processed nightsoil carries with it a high risk of spreading human disease. This is well evidenced by the high mortality rates in many Third World countries, and by the mild to severe intestinal problems that befall many travelers.

Thus, knowing the potential danger contained in humanure, I am not one to promote the use of human waste for root vegetables or greens that touch the ground. However, if compost with humanure is brought up to pasteurizing temperatures (160 degrees fahrenheit) repeatedly, then it may be safe. If you compile good raw moist ingredients in a compost pile all at once, the heat is likely to rise, within 24 hours, up to 170 degrees, but only in the interior of the pile. The closer you get to the outer surface, the lower the temperature. In those cooler places pathogens are not killed, and as you remove compost, the outer gets mixed with the inner, which makes it difficult to obtain pathogen-free compost from toilet wastes. I believe that if we are to use compost containing human bodywastes to grow food for humans, the compost must be brought to pasteurizing temperature throughout, repeatedly, in order to kill the various pathogens that can cause disease.

Pasteurizing temperatures can be accomplished through controlled rotating in-vessel compost processing (at great cost), where every ounce of the compost reaches 160 degrees or higher. Such high temperatures can also be reached with various heatingfuels, but this is also very costly, and causes pollution and depletes resources.

In addition, solar power can accomplish the job of safely pasteurizing compost, as I proved in the following experiment: I built a simple wooden box, about 3 feet long by 2 feet wide and 6 inches deep, and lined the bottom with one inch of rigid foam insulation. I put in finished toilet compost to a depth of 5 inches, and inserted two min/max thermometers. I made a double-glazed cover and fastened it securely, and at 11 a.m. tilted the box to face straight toward the sun. It was a bright sunny day, and by 2 p.m. it was 150 degrees inside. It of course cooled off at night. The next few days were also sunny, and each day, as the compost became drier and drier, the temperature rose still higher, until it reached over 170 degrees throughout the compost, more than enough to kill the toughest pathogens (see chart). With concentrating collectors far higher temperatures can be attained.

Thus we can produce excellent compost from human wastes that could be safe even for growing human food. For large municipal scale I recommend repeated pasteurization by slow conveyance through a long solarheated tunnel.

Unfortunately, this high-temperature solar-heat processing would kill not only the harmful organisms, but also the beneficial ones. Therefore I recommend pasteurization only in cases where the compost is truly needed for human food production. I prefer to use compost containing human wastes for ornamental plantings, and for production of animal food, fiber, fuel, lumber, paper, christmas trees and so forth. For human food we can use animal manures, since there seems to be little chance of disease transmittance to humans through the use of well composted manure from vegetarian animals.

FOOD PRODUCTION

❦ ❦ ❦

Living in my new home was a joy from the very beginning. I spent practically all my time there, because it also houses my weaving studio. I continued my profession as a weaver, with one or more resident apprentices, and soft, colorful shawls, blankets and rugs came rolling off the looms. We were basking in the sun, glorying in the home-grown tomatoes and salad greens and the fragrance of flowers, while the weather presented yet another blizzard.

When I realized that all the various solar-dynamic and bio-benign designs and methods in my home were working out far beyond my highest hopes, my mind increasingly filled with visions of sustainable life support systems for all, locally and globally.

Word started spreading about my home. More and more people wanted to visit, especially to see the year-round 30-foot tomato plants and the clean and odor-free composting toilets. Soon I felt the need for a place more accessible to the public, to demonstrate the wonders of good non-polluting, sustainable design. I became particularly interested in working further on the subject of food production: how can we - as families, communities, nations - best be assured of steady and plentiful supplies of nutritious and safe foods all year round? And how can this be done in ways that greatly reduce the harm caused by the current standard food production methods, such as harm to our water, air, and soil, and to our health, jobs and economy, and the waste of oil, minerals, water and other limited resources?

I knew about the drastic drop in aquifers and lakes around the nation and the world, and the vast areas laid to waste due to salt buildup, both caused primarily by modern sprinkler irrigation. I remembered the polluted water, too repulsive to even bathe in, in Iowa and in central Massachusetts, in those cases caused primarily by poor management of animal manure and by toxic pesticides.

I remembered the shocking experience of trying to pick up a handful of topsoil in central Massachusetts, where there was once the richest farmland. After many years of chemical fertilizers and toxic pesticides, heavy machinery and no compost, the soil was hard as concrete, void of organic substance or life force.

I also learned that the total energy consumed by U.S. agriculture per year is equivalent to more than 30 billion gallons of gasoline (714,285,000 barrels), and that this represents more than 5x the energy content of the food produced.

Thus, prior to the successful food growing experience in my home, I was among the many who had come to fear that the destructive modern food production methods, combined with the rising world population, would inevitably lead to our doom.

The Dream Greenhouse

I started envisioning commercial and community solar greenhouses on the Vineyard, requiring little or no fuel for heating or cooling, producing wholesome organic food for this community and beyond, and demonstrating the potential for increasing local self-reliance and decreasing dependence on food grown far away using destructive processes.

Around this time, a group of people interested in sustainable design gathered on the Vineyard for a slide presentation featuring Shane Smith's community solar greenhouse project in Cheyenne, Wyoming. Some of us were greatly inspired by this wonderful project and began talking seriously about starting a similar operation on the Vineyard. We proceeded to plan and design, but, unfortunately, the group leadership had no faith that such a greenhouse could work on the Vineyard (not enough sun), and thus the idea was quashed. (A couple of years later the group did build a greenhouse, but it was a standard double-poly one, heated with oil and cooled with two giant fans. In spite of this, they called it a "solar greenhouse". Such denial or lack of understanding of reality is unfortunately widespread and has severely limited the development of truly comprehensive solar design, to the great detriment of individuals, communities, and nations, and the general well-being of our planet.)

However, I continued to feel the urgent need to create a fully energy-self-sufficient greenhouse. Within a few days after I gave up hope that the group would join in such a project, I received a note from a friend with a clipping from Science Magazine: "A greenhouse grower in Oregon is reported to have cut more than two-thirds from his $1,000/month heating bill last winter by turning off his conventional heating system and installing 450 New Zealand rabbits... " My friend's (90 percent tongue-in-cheek) note to me was: "Do you think Vineyard rabbits could do as well?"

This set off an explosion of inspiration in my brain: how obvious! Warmblooded animals have been used for thousands of years to help keep people warm, housed in barns connected to human dwellings. Might it be possible, by combining animal and solar heat, to maintain an energy-efficient solar greenhouse not just barely above freezing, but warm enough for high-yield food production, maybe even tomatoes, through the coldest New England blizzard, without using any heating fuels?

I immediately set about to design such a greenhouse, researching and calculating every detail. I found out that the heat output from both rabbits and chickens is 8 BTUs per hour per pound of body weight (one BTU, or British Thermal Unit, is the energy required to raise the temperature of one pound of water by one degree). Thus for the six cold months the heat production from each animal would be equivalent to 2 to 3 gallons of heating oil.

I decided to dedicate the least well-lighted spaces of the greenhouse to the animals, 100 chickens in the northeast corner and 30 rabbits in the northwest. I calculated that they would pay for their feed with the eggs and angora fiber they would produce, and the warmth and compost fertilizer would be free.

It wasn't until considerably later that I found out about yet another extraordinary benefit the animals would provide: the carbon dioxide that they and their bedding exhale. I learned that in places like Holland, growers had discovered that by enriching the air in greenhouses with CO_2, released from tanks, they could greatly increase the rate of plant productivity. It was reported that this CO_2 enrichment was as costly as the heating bill. Could the animals possibly provide effective CO_2 enrichment for free?

A few people at New Alchemy Institute agreed that incorporating warmblooded animals seemed like a logical next step, but most people and printed materials said it could not be done. Statements claimed it would be too hot for animals in a greenhouse, they would eat the crops, create dust, and something vague about foul air, but no one knew specifics. And there was also the same old objection I heard so often: "If it could be done, people would already be doing it, and since no one is doing it, it must be impossible."

One of the most important questions for this greenhouse project was: which is the best available glazing material? I did not want the usual greenhouse glazing, which had only a three-year life span and then had to be discarded. Someone at New Alchemy Institute recommended I call Gary Krogseng at 3M Company in Minnesota. Gary described the superior light transmittance (97 percent), UV resistance, strength and longevity (10 to 15 years) of the Sungain glazing material, and then he listened carefully to my proposed innovative greenhouse. He said that the Sungain glazing was originally developed to be placed between two layers of glass, to form tri-pane or quad-pane windows. 3M had given Gary permission to try it on an experimental basis as a stand-alone glazing for greenhouses. Two small projects were in place and he was looking for a third, in New England. Would I be interested in being the third experiment and receiving the glazing free from 3M?

Immensely grateful and energized by this offer, I proceeded full speed with the designs. After months of calculating, drawing and redrawing, and meditating to seek the highest and innermost guidance, I had finalized every detail to the point that it coordinated with every other detail in a mutually harmonious, logical blend.

Another major question: how was I to pay for this experiment?? I looked into grants but was discouraged by the memory of the shallow and rejecting response I had received when I applied for a government grant to build my home a few years earlier. Clearly, the reviewers of the grant did not understand what I thought was, and what later proved to be, possible to accomplish.

Believing that we human beings have no time to lose if we are to avoid global disaster, I decided I must try to put into reality this dream of a greenhouse capable of growing high yields of high-quality food without fuel or pesticides throughout the worst New England winter. With the help of a sympathetic local banker and with my land as security, I borrowed money from the bank. As the project progressed I also received generous donations and loans from people who understood the significance of the project.

I couldn't possibly afford to pay for all the labor, so my plan was to hire one professional carpenter with the tools and skills needed to build a sound structure and to seek volunteers for the remaining work force. I sought these volunteers in a radical way that I recommend to others who plan an innovative project for the general good of humanity and have the housing to support it: I invited the world to come join me in constructing this greenhouse project.

I made a drawing of the greenhouse dream, photographed it, and put copies on fliers describing the project. I offered room and board in exchange for work. I sent the fliers to various institutions and colleges that I knew had strong commitment to environment and future. Within days I got the first response: my New Alchemy friend Robert "Sardo" Sardinsky, photographer, writer and expert on solar electricity, said he would come, and he would bring some friends.

Actualizing the Dream

It wasn't until late July that finally I dared to take the terrifying plunge and begin this 4,000-square-foot project. On day one, July 27, 1983, I planted over 100 tomato seeds, to become the main crop in the greenhouse. The same day a bulldozer broke ground, and then concrete footings and foundation posts were poured.

On August 16 we raised the first rafters, and people came flocking from all over to volunteer. From Australia, Germany, Canada, Denmark and Sweden, from Vermont, Maine, New York, Maryland and California, some were highly skilled carpenters, while others had no experience but quickly learned. I made sure materials were always on hand for the next stage of the project, and that there was plenty of food in the house.

People were up in the sky joining rafters, hammering and sawing, lifting, carrying and digging - all the while singing, joking and laughing. The greenhouse progressed rapidly and the site was immaculate, mainly because one highly skilled and efficient angel, Jonathan Schuall, set the tone by keeping everything sorted and organized. In addition, people tended and harvested the bounty of the summer garden, and vied for their turn to cook. We were 8 to 10 around the table, starting each meal by holding hands in silent meditation, giving thanks. Nourished by lovingly prepared food, we shared our experiences of Great Co-Incidences and other inspiring personal stories. We all agreed we had never eaten better or ever been part of a more unifying, satisfying community experience.

The land on which the greenhouse was rising, previously part of a dairy farm, had for 70 years been a pasture and hayfield that had never received any toxic pesticides or chemical fertilizers, only composted cow manure and lime. It had 10 inches of good loam topsoil with a pH value of 6, sandy subsoil and good drainage, and was relatively free of stones, at least the kind that break rototillers. As the framing neared completion, the ground inside was layered with aged black peat from the bottom of a pond, brown Canadian peat moss, compost from a nearby farm, lime, greensand, and soft rock phosphate. We rototilled it all in, three times over. I wanted to spare no expense on soil preparation, because that is the foundation of the entire production.

Next, we removed the newly blended soil from the northeast and northwest corners, where the chickens and rabbits would live, and we erected the wooden framework for the first-level waterwalls, to serve the dual purpose of separating the chickens and rabbits from the plant area and to hold the mass to absorb and store the solar and animal heat.

Then we built the growing beds. I wanted them to be high enough and narrow enough to be comfortable to work with. We placed strings to mark paths and dug the prepared soil down to subsoil level to create the path areas and piled it high in the center of what was to be the growing beds. We put in cedar posts along the sides of the beds and stapled reenforced plastic fabric on the outside of the posts. Then we nailed spruce fence boards to the posts. The plan was that the plastic would prevent the fence boards from rotting. It worked, as I will explain later in the section on construction details and recommendations.

Seven weeks after groundbreaking, the framing and most of the interior were completed, and the south rafters were painted gleaming white. Right on schedule, fall equinox 1983, Gary from 3M descended from the sky in a brilliant sunset, landing at the little Vineyard airport. He had arrived to help install the glazing. The next morning was not the calm day we had hoped for to facilitate the painstaking work of applying the glazing. Instead there was a whipping wind. Another task also became top priority that same day, which further complicated the glazing plans: we had to plant the tomatoes. The 100-plus tomato seeds that I had sown on day one, July 27, had become 100 thriving young tomato plants 18 inches tall. They were growing in half-gallon milk cartons on the deck up at the house, waiting patiently to be planted into the new greenhouse, and this day they were getting beaten and blown over by the wind. It was now or never for them.

So it came to be that this marvelous group of people worked all day with total dedication as hard as any of us had ever worked. The nurturing women carried the young tomato plants like babies across 800 feet of windy pastures and tucked them into the beds in the greenhouse, while the brave men were high in the rafters wrestling to apply the invisible but noisy plastic glazing with special double-stick tape.

That night Gary took us all out to dinner to celebrate, and afterwards we went to play under the full moon, inventing games and running around like kids, first at the beach and then on the highest hilltop on the Vineyard. Nothing beats collaborating successfully on a project for the good of all.

There were to be four layers of glazing on the whole 2,400-square-foot south facade of the greenhouse, and in three days we had applied layer No.1 on the outside of the white rafters and layer No.2 on the inside. In addition, we had applied layers No.3 and No.4 on several sections. Before returning to Minnesota, Gary set up a special light meter capable of simultaneously reading the light levels inside and outside. The meter showed an astonishing 95 percent light transmittance through four layers of Sungain glazing. This was considerably more than what just one layer of glass or polyethylene transmits. We also installed a 600-gallon galvanized steel farm water tank, and connected it to a 1,000-foot black poly pipe for solar-heating the water. We also created a fire chamber underneath this tub, because I wanted a backup heat source, just in case.

In mid-October my brother arrived from Sweden to install the C.I.T. (Capillary Irrigation Tubes) underground irrigation system he had invented several years earlier. These special formula low-fired ceramic pipes, which can be made from local clay anywhere in the world, had already been installed in arid areas such as Botswana, Egypt, Peru, Mexico and Sri Lanka, providing astonishing increases in yields and reduction in irrigation water. Also tested in Sweden, this system has met with equal success, growing cut-and-come-again shrub willow for the production of fuelwood. Buried permanently below ploughing depth, the C.I.T. system can remain effective without maintenance essentially "forever". The test installations have demonstrated that this is the most economical, effective, water-conserving and low-energy irrigation system in the world. Solviva would provide the opportunity to test this system in a northern greenhouse.

It took two days to install 400 feet of C.I.T. pipe 12 inches deep. To test it, we put a hose into one of the 13 upright vent pipes distributed along the system, and after a few minutes the water level started rising in all the other vent pipes, confirming that the whole system was evenly filled with water. We then turned down the flow of water from the hose and left it on for 12 hours. The water level in the vent pipes remained at the same level, demonstrating that the water was being absorbed by the soil surrounding the clay pipes.

By Thanksgiving the weather was turning cold and the greenhouse was almost all closed in and insulated. All four layers of glazing had been installed on the lower half of the greenhouse, but the upper half still had only two layers. Thousands of salad plants and herbs were growing in the raised beds along with the 100 tomato plants which were by now about 6 feet tall and loaded with many hundreds of green tomatoes.

The photovoltaic panels, batteries and related equipment were installed by Sardo, to power the pumps and fans to circulate the solar-heated water and air to their respective storage places. When Sardo had finished mounting all the delicate controls on a framed board by the east entrance, he wanted to protect them from moisture and dust. He wondered if perhaps I had something more attractive and longer lasting than plastic to cover the master control panel. I recalled an old stained-glass window that I had put into deep storage some 10 years earlier. After much rummaging I found it, and when I brought it to him, we discovered to our astonishment that it was exactly the same size in both height and width as the frame he had just built, down to 1/16 of an inch. This was one of the more powerful of the many Co-Incidences (happening with God) that again and again graced the Solviva project, like an encouraging pat on the back from God. There is definitely SOMETHING going on.

Animal Power

One hundred little day-old fluff-ball chicks had arrived by mail in mid-September. By Thanksgiving they were half grown and installed in their spacious and cozy quarters behind the waterwall in the northeast corner of the greenhouse. A thick layer of leaves covered the dirt floor, providing the chickens with endless entertainment, cover for their droppings, and dust baths. Automatic devices dispensed a continuous supply of feed and fresh water.

Three gorgeous fluffy gray-brown angora rabbits were brought down from Maine by Sardo, one buck and two does. While we built their enclosures behind the waterwall in the northwest corner of the greenhouse, the rabbits were temporarily housed with the chickens. They all got along surprisingly well, but I think the rabbits were happy to finally have apartments of their own, no longer having to put up with being continuously groomed by the chickens.

The three sheep that I acquired in 1981 had by now multiplied to twelve. They moved into the spacious barn built against the slanted back wall of the greenhouse, and we piled their winter's supply of baled hay up against this wall to further increase the greenhouse insulation.

The First Winter

Then the "Canadian Express" arrived with a roar a few days before Christmas. Deep snow, howling winds, temperatures plunging below zero degrees F. We quickly found out where there were still holes and incomplete insulation. My daughters and their friends arrived for the holidays, and the first tomatoes ripened, as sweet and delicious as the famous ones in my home.

On sunny days the greenhouse was as sweet as the tropics. The solar heat warmed the waterwalls to 85 degrees F. The 600 gallons of water in the big tub was warmed to 100 degrees by the sun shining on the 1,000 feet of black poly pipe in the peak of the greenhouse. The water in 350 jugs, set in an underground tunnel, was warmed to 75 degrees by the hot air ducted down from the top of the greenhouse.

On the last night of 1983 it was bitter cold, and there had been no sunshine for a couple of days; and because the greenhouse still was not finished, the low temperature threatened to damage the tomatoes. So we lighted a roaring fire of scrap wood, cardboard and newspaper in the fire chamber under the big tub. The hot stovepipe quickly warmed up the whole greenhouse, and after about two hours the water in the tub had risen from 55 degrees F to over 100 degrees.

It did not take long for me and my three daughters to recognize a great opportunity. We all got in, fitting easily up to our chins in the 7-foot-diameter, 2-foot-deep tub. We tossed in fragrant herbs and oils, and soaked and hummed in the hot water for hours in candle light. This was bliss... free-floating in hot water, in a thriving garden, exquisite tomatoes within reach, with cozy happy animals to the east, west and north, while dark cold winter roared outside.

When we finally were ready to leave the tub, we discovered that one of the female rabbits had given birth. Deep down in the coziest imaginable angora-lined nest lay eight little silky blind bunnies, the size of my thumb, three gray-brown, two black, two white and one golden.

A few weeks later we applied the final layers of glazing to the entire south facade, completed the insulation, and plugged up the remaining gaps. Backup heat was no longer required.

The Greenhouse Cornucopia

Tomatoes were now ripening by the hundreds, with the same superb flavor and texture as the ones in my home. Herbs and greens grew extraordinarily lush and delicious. It was time to begin marketing Solviva produce.

I couldn't bear the thought of decapitating the heads of lettuces and other greens in their prime, so I started to harvest individual leaves of the 30 to 40 different varieties of salad plants and herbs. I washed and drained them gently, then packed them in plastic bags in half-pound lots topped off with edible flowers like nasturtium and alyssum. I wondered how long their shelf life might be, and whether people would be interested in buying salad this way. I brought samples of this mixed clean salad to chefs, who had never seen anything like it before. (Ready-to-serve salad blends were still an unknown concept at that time.) They soon realized that the flavor, beauty and convenience of these salad greens was far superior to anything else they could buy, and orders began to come in.

At that same time some troubling news became top priority in the media: EDB (ethylene dibromide) had been used extensively in American food production, for both growing and storing grains, vegetables and fruits, because it reduced damage by various insects. But Governor Graham of Florida had been presented with persuasive scientific evidence that EDB was a powerful carcinogen, and in early January 1984 he blew the warning whistle and ordered the removal of many food items from stores.

Soon other governors around the U.S. followed suit. Governor Dukakis of Massachusetts ordered the posting of proclamations in all food stores, with a long list of affected foods and their EDB count, advising everyone to buy organically grown food until the EDB-laced food was purged from the shelves. This news was more effective for marketing the Solviva produce than a million-dollar ad campaign could have been, and both the Solviva Salad and the tomatoes flew out of the stores as quickly as I could supply them.

Trouble in Paradise

Blotches appeared on the tomato leaves - first dozens, then hundreds, then thousands. Pruning off all the sickly looking leaves was decreasing the plants' ability to photosynthesize. Could it be a nutrient deficiency or excess, or bacteria, virus or fungus? I searched through all my books and bought a few more, including a big British volume called TOMATOES, but I found no likely explanation. I called the Massachusetts Extension Service and departments of agriculture, horticulture, plant pathology at the University of Massachusetts and other institutions. Finally one older professor at U Mass asked in detail about the management of the animals: "Does it ever smell of ammonia in the greenhouse?" Yes, sometimes, because, in spite of my efforts, some of the nitrogen in the animal manure was evaporating as ammonia.

I certainly knew that ammonia was not pleasant for people to breathe, but it turned out that just 14 parts per million of ammonia in the air, barely detectable by our noses, could be harmful to sensitive plants, tomatoes among them. The ammonia causes burns to the stomata (nostrils) on their leaves.

In order to prevent ammonia from entering the plant area, I sealed in the animal areas with plastic, and covered the floor with more and more carbon material in the form of sawdust and leaves to absorb the nitrogen and prevent its transformation to ammonia. But this was still not enough. Monitors for reading ammonia levels in the air were available, at hundreds of dollars, and filters for capturing the ammonia cost even more, and both would require time-consuming maintenance.

Someone offered zeolite as a possible solution and sent me a package. I sprinkled the white powder on the bedding, as suggested, and it worked well, apparently binding the nitrogen. But this used up a lot of zeolite, and I did not like the idea of becoming dependent on this quite expensive volcanic substance which was mined in the deserts of Arizona - especially after I had a chance to experience those deserts and their exquisite, delicate beauty.

A friend made a magnificent attempt at constructing an air filter that would use far less zeolite, and we were told that we could actually reuse the initial zeolite after soaking the filter in water, thus dissolving the trapped ammonia and transforming it into nitrate. This nitrate-enriched water could then be used for supplying essential nitrogen fertilizer to the roots of the plants. However, this filter proved cumbersome to manage, and there was still some ammonia damage.

I was beginning to fear that this idea of having animals in the greenhouse was perhaps impossible after all. Yet, it was clearly so right for so many reasons. A New Alchemy friend brought over a CO_2 measuring device, and the reading among the plants was around 1,400 ppm of CO_2, four times more than the normal ambient of 350 ppm. Coincidentally, this was exactly the amount that the Dutch were adding to their greenhouses, at great expense. It was obvious from the growth rates in the Solviva greenhouse that the plants were far more productive because of the CO_2 enrichment provided by the animals. We simply had to find a way to keep the animals in the greenhouse without causing ammonia damage. And after two winters of discouraging struggle we did find a way.

Solution: The Earthlung Biocarbon Air-Purification Filter

Bruce Fulford had learned in Holland about ways to use compost to heat a greenhouse, and at New Alchemy Institute he designed and built a small greenhouse that incorporated these techniques. Tons of bedding from stables and dairies were loaded into a long chamber built against the whole north wall. The composting process generated heat for a few weeks, and then the old batch had to be replaced with a new one. Fans circulated the warm air from within the compost bank through ducts that ran through the growing beds within the plant room. The warmth and CO_2

benefited the plants, while the harmful ammonia was trapped in the soil and transformed into beneficial nitrate. Together we adapted the Dutch design for the Solviva greenhouse and constructed a filter system to scrub the ammonia out of the air coming from the animal quarters before it enters the plant area.

We dug the topsoil out of the bed next to the chicken room and lined the sides of the bed with airtight plastic fabric. Then we put in coarse gravel, and on top of this laid down perforated pipes, capped at the ends. This was topped with more coarse gravel. We covered the gravel with porous nonrotting landscape fabric, and then topped this off with a 12-inch layer of leaf mold mixed with sandy soil. We hung four branches of flexible ducts within the chicken room, and joined them into one duct containing a small DC fan powered by the sun through the photovoltaic panels and batteries. Then we connected that duct with the perforated pipes laid in the gravel. Finally we seeded this bed so plants could absorb the nitrate.

Thus, the air from within the chicken room, containing ammonia (harmful) and CO_2 (beneficial), is blown through the gravel layer and then rises to the only place it can go, up through the leaf mold/sand layer. This layer is permeated with tunnels created by earthworms, pill bugs, centipedes, millipedes and earwigs, as well as myriads of microscopic flora and fauna. In these dark, humid passages the moisture and bacteria capture the ammonia and transform it into nitrate. The nitrate is then absorbed by the microscopic root hairs of the plants that grow in the filter bed and then propelled up the stem to perform miracles in every part of the plants' bodies.

The CO_2 molecules from the breath of the chickens and the compost pass through the tunnels and passages in the Earthlung filter in the only direction they can go, up through the surface of the Earthlung bed, and then waft through the air among the plants. There the CO_2 molecules are inhaled by the plants through the stomata on their leaves, and then they flow through the veins of the plants, providing the essential building blocks that enable the plants to grow profusely.

In a conventional greenhouse CO_2 can be so depleted by 11 a.m. that the plants' ability to grow is greatly reduced. No matter how well other requirements are fulfilled, such as temperature, light, water, and fertility, if there is not sufficient CO_2, growth stops. Conversely, if CO_2 increases above normal, growth also increases.

Thus, in one simple, elegant, economical, low-maintenance design this Earthlung filter turns harmful ammonia into beneficial nitrate to feed plants through their roots and at the same time distributes the CO_2 in the air to feed the plants through the stomata on their leaves.

Birth of Solviva Salad

Over the first couple of years I grew well over 150 different varieties of plants inside the greenhouse. The tomatoes were the dominant crop at first, with salad greens, cooking greens, various root and fruit vegetables, and herbs secondary. At first I was selling the produce at standard produce rates, but the ratio of work and expenses to income was not encouraging. I needed help to find ways to increase the income.

I learned that Gus Schumacher, then Commissioner of Food and Agriculture in Massachusetts, was actively promoting local food production, so I made an appointment to meet with him. On a cold winter day I took the ferry across Vineyard Sound and traveled to Boston, and presented him with a bag of Solviva Salad topped off with a few nasturtiums and other edible flowers. He immediately started sampling the contents and passed it around to his staff. All exclaimed about the superior flavor, texture and appearance. He was amazed when I told him that throughout that bitter cold winter I had been growing this in a greenhouse without any heating fuel, lights or pesticides. "Clearly this proves that Massachusetts could become salad self-sufficient year-round. You just need some help with marketing to prove that it can also be a good business."

At that time Gus Schumacher was organizing banquets at some of the finest restaurants in Boston, to promote "Massachusetts Grown and Better" and to honor Massachusetts growers and the chefs who use their products. He wanted to buy Solviva Salad and edible flowers for 1,000 guests for the next event, three weeks hence. Could I supply? Could I supply! A 1-ounce salad serving for 1,000 guests amounts to about 63 pounds, plus 2,000 flowers. No problem.

On the appointed day, dressed in a long silk gown, I again crossed the water and drove up with 13 large boxes. It was indeed a glittering event, and the response to the salad was amazement and excitement. Orders from Boston-area chefs started to pour in, more than I could fill.

I wondered how production within the Solviva greenhouse could be increased. Where else could salad plants grow besides the existing raised beds? Perhaps there could be hanging gardens above the beds, perhaps in aluminum gutters. The aluminum is anodized and therefore unlikely to leach aluminum into the produce, but there were two major questions: 1) would the gutters offer enough root space for high yields and prolonged production, and 2) would the hanging gardens reduce productivity in the raised beds below?

At first I intended these hanging gardens to be hydroponic: the gutters would be filled with perlite, with nutrients supplied in solution. I did not want to use the usual hydroponic nutrient mix containing only a few chemical nutrients, because I felt this contained no life force and would supply far fewer varieties of nutrients compared to what compost provides. So I planned to make compost tea by steeping a bag of compost in a barrel of water and to periodically pump this nutrient-rich water into the gutters. But when I suspended the compost in water, it became immediately clear that this was not to the liking of the earthworms or any of the other aerobic organisms in the compost. So I decided against hydroponic culture and instead filled the gutters with compost mixed with peat moss, perlite and vermiculite.

We punctured drainage holes in the gutters, filled them with the compost mix and young lettuce plants. We then suspended two gutters side by side about 2 feet above the wide growing beds. The lettuce plants grew phenomenally fast and lush, and, with weekly supplementary feeding with liquid seaweed, and leaf-by-leaf harvesting once or twice a week, they remained highly productive for over two months. And there was no discernible reduction in productivity in the bed below.

Because of the success with these growtubes, I wanted to hang up more, but did not like the idea of buying more of these gutters, because aluminum requires enormous amounts of electricity to produce. So I substituted 4-inch-diameter PVC plumbing pipe, which also had the advantage of being half as costly. We taped the ends, drilled drainage holes on the bottom, cut out pockets on the top, and filled the new growtubes with compost mix and young plants. To my delight, even though the root space was even smaller than in the aluminum gutters, the production was just as good.

Then we added a second tier of growtubes one foot above the first, and soon, because there was still no significant reduction in growth in most of the plants in the beds below, a third tier went up, and finally a fourth. Thus eight growtubes were suspended over each of the 10 major beds, for a total of 80. Before long we also installed three tiers along the whole length of the upper catwalk. All in all, we managed to fill every available nook in the greenhouse with a total of 125 growtubes. This just about doubled the available growing space, and production soared to as high as 90 pounds some weeks, or 1,500 servings. This was considered extraordinary for a 3,000-square-foot greenhouse.

Gradually over the years, through observations, experiments and experiences gathered while managing the indoor and outdoor gardens, I developed a comprehensive system that proved capable of producing exceptionally high yields of high-quality food year-round. These Solviva management techniques work in harmony with the laws of Nature. In order to have even a chance at success with a commercial operation, it is of utmost importance to understand these laws and to follow them in accordance with your best judgement. What I can say with certainty is this: your level of productivity and profitability is directly related to how happy the plants are and how efficiently production and business are managed. Needless to say, a home greenhouse to provide for family and friends can be run ever so much more casually and still be absolutely wonderful.

Later on I will describe in detail the methods that resulted in such exceptionally high yields of the highest quality food year-round from a very small piece of land, without any pesticides, heating fuels or cooling fans. But now that I have described how I manage heating and cooling, wastewater, and food production in ways that minimize harm, I want to relate some of my experience with electricity, transportation and solid waste management.

❧ ❧ ❧

ELECTRICITY

The average household in New England consumes 9,000 kilowatt-hours (kwh) per year, not counting electric heating. At a rate of $0.15 per kwh (current price in this area), this amounts to $1,350 per year. This seems a reasonable price for something as important as electicity. However, this electric bill represents only a small portion of the true cost of producing electricity with today's conventional technologies, oil, coal, nuclear or large-scale hydro, what Amory Lovins calls "hard path" technologies.

Some of the unbilled costs of conventional electricity production are built into our taxes, in the form of many billions of dollars in subsidies and tax breaks for the oil, coal, hydro (big dams) and nuclear industries. Other costs, such as nuclear wastes and groundwater pollution, are being deferred to burden future generations. Many more costs are not reflected in either electric bills or taxes, but are hidden in our society in many ways. In order to know what we are really paying for our electricity, we must obviously include all these real costs.

For instance, we must include the real cost of the damage caused by the acid rain that results when power is generated by coal and high-sulphur oil. Acid rain poisons, kills and reduces the health and productivity of forests, fields and lakes. It causes bridges and cars to rust, paint to blister and peel off buildings, and stone buildings, monuments and ancient temples to crumble.

We must include the real cost of the loss of millions of acres of land flooded by huge hydroelectric dam projects, such as Canada's HydroQuebec, and vast areas laid to waste by mining, tailing, and severe pollution.

We must include the real cost of the military force that is required to combat and prevent the terrorism and wars that result from competition and dependency on foreign oil, and the promulgation of bomb-grade radioactive materials resulting from nuclear power plants. How much of the military is in place primarily to protect our continuing and increasing access to oil in foreign lands? (According to War Resisters' League, 50 percent of our taxes is spent on current and past military expenditures.)

We must also include the real cost of the many health problems that result from the widespread pollution of air and water that is caused by conventional electricity production.

Experts say that when these true costs are included, the actual cost of producing electricity with today's conventional technologies is many times higher than our electric bills. The bottom line is much farther down.

But there are better technologies available right now: sustainable, renewable and far less polluting, as well as safer and more dependable and secure - and, in actuality, far less costly. These are what Amory Lovins calls "soft path" technologies, which include efficiency and various renewable technologies powered by sun, wind, and water, as well as plant-based energy sources, such as wood, methane and alcohol.

Not much controversy remains about efficiency: it is clear that today's best motors, pumps, and light bulbs can get the job done with some 75 percent less energy than conventional technologies. Thus several electric companies are practicing "negawatt" (again, thanks to Amory Lovins) policies, providing energy-efficient refrigerators and light bulbs to their customers at far less than the cost of a new dam or nuclear power plant.

But there is a widespread belief that renewable technologies are impractical, unreliable and far too costly - and purveyors of "hard path" methods seem to have a vested interest in maintaining that impression.

Solar Electricity

Of the various renewable technologies available today for making electricity, photovoltaic, or PV, is my first choice. Many people use small amounts of PV power in their daily lives nowadays, calculators and watches, and - less common but nonetheless widespread - battery rechargers, exterior lighting, flashlights, cooling fans. But for normal electricity requirements most people believe that PV technology is far too inefficient, impractical and uneconomical to be even worthy of consideration at this time.

Photovoltaic electricity technology was originally developed by the space program. PV cells are made from silicon, a major component of ordinary sand, and they can be manufactured in ways that are far less polluting than current methods. To greatly simplify: photovoltaic cells consist of thin layers of silicon, one with a positive charge, the other negative, with a gridwork of metal conductors in between. Electricity is generated when sunlight causes a charge between negative and positive layers. The panels are silent and seemingly motionless, but on the molecular level they are dancing and humming, capable of producing electricity reliably for 20, 30, 50 years or more, requiring hardly any maintenance. The price of PV panels has fallen dramatically over the years and will continue to do so as mass production and demand increase.

Many have the impression that PV is too inefficient to be practical, but this is a grave misconception. PV is plenty efficient right now, capable of producing around 10 watts of electricity per square foot. In a sunny southern place such as Arizona, this rate of production amounts to 22 kwh per year per square foot of PV panel. But what about cloudy New England? One square foot of PV panel generates about 15 kwh per year here in New England. An efficiency increase of 20 percent can be realized if more sunlight can be reflected onto the panels from an adjacent pond or other reflective surface. That way each square foot of PV panel can generate about 18 kwh per year in New England, about 27 kwh in the sunshine states. The panels can also be mounted to track the path of the sun, thereby increasing efficiency by 25 percent.

As mentioned earlier, the average household, without electric heating, consumes 9,000 kwh per year, which at 15 cents per kwh comes to $1,350. In New England, 500 square feet of PV panels, installed with reflective device, are required to generate 9,000 kwh per year. The current price of PV panels is about $64 per square foot, so 500 square feet would cost about $32,000, or $36,000 with installation. Amortized at 8 percent this comes to $2,880 per year, clearly not cost-effective compared to the superficial bill provided by the power company.

However, electricity consumption can be reduced by more than 65 percent or more with various efficiency technologies, without any change in lifestyle. The more heat that is produced by an appliance, tool or light bulb, the more electricity is required to run it. For instance, ordinary lightbulbs get very hot and therefore consume a lot of electricity. A family may have ten 75-watt lightbulbs lighted an average of five hours a day. That is 3,759 watts, or 3.76 kwh, per day. Multiplied by 365 days per year, this amounts to about 1,372 kwh/year, which at $0.15/kwh costs $206 per year. By contrast, the best compact CFL (compact fluorescent lamp) provide the same amount of light, but consumes about 75 percent less power because they hardly even get warm. Therefore, ten of these energy-efficient light bulbs consume only 343 kwh per year, costing $51 @ $0.15, saving $154 per year. The bulbs cost more to buy, but because of their energy efficiency and much longer life span (10,000 hours versus 1,000 hours for regular light bulbs), they provide a tremendous return on the investment, risk-free. Similar cost-effective efficiency is available for refrigerators, washing machines and other appliances.

Thus, by taking comprehensive energy-conserving measures, an average home can reduce electricity consumption from about 9,000 kwh per year to 3,000 kwh, without sacrificing comfort or convenience. This would require only 166 square feet of PV panels, with reflective device, covering an area about 8 by 21 feet. At $64 per square foot, the PV panels would then cost $10,624, say, $13,000 including the cost of installation and inverter, or about $1,040 per year amortized at 8 percent. This is less than the normal household electric bill of $1,350, plus you gain a cleaner conscience because you have greatly reduced your contribution to global warming, acid rain, nuclear wastes, and depletion of resources. (The economic comparison becomes even more favorable when we factor in two items: that the actual cost of conventional electricity is much higher than what is charged on the bill, and that the price of PV panels will plummet when they get mass produced.)

This is the approximate economic reality today if you are connected with the power company grid and install a synchronous inverter and controls between your PV array and your electric meter. In this case the PV power runs back into the grid, actually running your meter backwards. On sunny days your photovoltaic panels will produce more than you need, and the excess will be sent to the power company. On cloudy days and for the 18 hours a day that the PV panels are ineffective, you will receive electricity from the power company.

ComElectric, the power company in this region, accepts properly converted private PV power into their grid, as much electricity as is being consumed, but no more. Thus, if you produce as much as you use, you pay only a small monthly hookup charge.

The cost is considerably higher if you choose to be independent of the electric company grid. Not only do you have to add and maintain batteries and a whole setup of different controls, but you also have to add extra generating capacity to provide power during extended cloudy periods.

There are two options for this. The more costly is to add more PV panels and batteries. A less costly option is to add a generator as backup. This generator can be powered by gasoline, or better yet, by a renewable resource such as methane gas produced from anaerobic digestion of wastes.

Going independent is cost-effective only if your site is located so far from the nearest power line that the expense of the electric cable and the trenching would be unusually high. Independent electricity production used to be cost-effective only if the distance between your site and the nearest grid connecting point was 10 miles or more. Then, as the price of PV cells went down and the price of trenching and cables went up, the distance was reduced to five miles, two miles, half a mile. Nowadays, independent PV electricity production is said to be cost-effective if you have to go further than one third or one quarter of a mile to hook into the power grid.

Whatever system it is, it must be as automatic, reliable and low-maintenance as possible. Like most women, I have no interest in fiddling with machinery. But I love the 4-by-6-foot PV panel array that was installed to provide the power for the pumps and the fans in the Solviva greenhouse, and the maintenance requirements are minor: it takes but a couple of minutes to slightly shift the angle of the panels a few times a year to be in the best relationship with the sun and to wipe off snow, plus 30 minutes twice a year to clean the battery terminals and top off the water in the batteries.

It is clear that solar electricity can make sense for rural homes. But it is generally considered utterly impossible and impractical for urban areas, that thousands of acres would need to be acquired and cleared for photovoltaic panels, and that the cost of panels and land would make it totally uneconomical, and besides - what about when there is no sun?

I am strongly opposed to taking thousands of acres of land and covering them with photovoltaic arrays, opposed for financial, aesthetic, social, and environmental reasons. But there is no need to do so because the required area is already available, on many sites within and around cities. In my favorite "clean/green, light/right, sustainable, solar-dynamic, bio-benign living" scenario, the electric company provides, installs and maintains grid-connected PV panels on thousands of sites. Any south-facing wall, roof, fence, embankment, lawn or slope is a potential site. PV can be installed along the highways, on fencing and retaining walls. High-rise buildings can have PV on the roof and between the windows on the south wall. PV panels can be installed over parking lots, thereby also providing welcome shade in the summer. Furthermore, PV panels can even be combined with space and water heating.

PV power would have minor operating costs compared to oil, coal or nuclear power, and almost none of the risks of accidents or terrorism.

Enough PV should be installed to supply not only for the needs during times of sunlight, but also enough to store in various forms of batteries. One form of battery that I imagine possible, a hydrobattery, consists of two water tanks, one above ground, the other in-ground next to it. While the sun is shining, PV panels pump water up from the lower tank into the upper tank. When there is no sun, the water is released through a series of electricity-generating turbines that lead into the lower tank. This form of hydropower would not consume any water, because the same water recirculates ad infinitum. The in-ground tank can be eliminated if the tower is built next to a river, lake or ocean. (I would appreciate a note from anyone who has heard of a similar installation, or who knows how much electricity could be produced by such hydropower.)

The photovoltaic electricity would be supplemented with a mix of other renewable methods of electricity generation. These combined renewable methods would in turn be supplemented with the usual mix of "hard path" technologies, but the consumption of oil, coal, and gas could be reduced by more than 80 percent, and nuclear power could be eliminated entirely.

Incidentally, for those who live in areas that experience periodic power outages (and who doesn't in these days of increasingly destructive weather?) I recommend a small PV setup, battery bank and supplementary generator to provide electricity during such power outages. It may be charming to light candles during a brief power outage, but the mood soon turns sour when the blackout continues for a week after a hurricane, as it did on the Vineyard a few years ago when Hurricane Bob came roaring through. People could not flush their toilets, take showers, do laundry, use computers, or watch television, and the food in thousands of refrigerators and freezers spoiled. An independent backup power system is enormously satisfying and money-saving during such times. It costs only a couple of thousand dollars and provides an excellent return on investment if we account for the real losses that occur during such prolonged blackouts.

Design for a Home-Scale Solar Power Plant.

This garage is sheathed with 482 square feet of Photovoltaic Panels on the south-facing roof and wall. The small pond increases the electricity production by 15-20 percent by reflecting more sunlight onto the P.V. panels. This installation is capable of producing over 8000 kwh per year (in New England!), more than enough for an energy-efficient home AND two electric cars.

The P.V. panels would cost $30,848 at today's cost of $64 per square foot, roughly $33,000 with the inverter required to phase the electricity into the power company grid, and a small battery bank and controls to provide the home with electricity during periodic power grid outages. Amortized at 8% this comes to $2,640 per year. By comparison, the electric bill for a normal home in New England is $1,350 per year, and the operation and repairs of two gasoline-driven cars is about $2,000, totalling $3,350. Thus the SAVINGS for this P.V. powered home and transportation would be roughly $710 annually.

This garage also has a 330-square-foot office/studio/apartment upstairs, heated and cooled by the sun, and with nonpolluting Biocarbon wastewater purification system.

TRANSPORTATION

❦ ❦ ❦

The transportation sector currently consumes over 25 percent of the energy used in the United States, almost two thirds of all the oil. Of that oil, more than half is consumed by the private automobile. Indeed, American cars consume one ninth of all the oil consumed in the world every day.

I have long believed that electric cars are far less polluting and more economical than gasoline-powered cars - provided the batteries are powered by the sun. For one, electric motors and batteries are much simpler to build and to maintain than the internal combustion engines. For instance, electric motors do not require oil changes and radiator flushes. They never need tune-ups, antifreeze, or carburetor adjustments. They have no mufflers, spark plugs, valves, fuel pumps, fan belts, water pumps, pistons, radiators, timing belts, condensers, points, or starters. They have no need for PVC valves or catalytic converters and never need smog tests.

If electric motors, photovoltaic panels, batteries and the car bodies are manufactured in ways that cause 90 percent less pollution than current practices and are 90 percent recycled, both of which are doable now, and if the energy to power the batteries comes from the sun, with backup provided by other sustainable technologies, with perhaps a 10 percent backup provided by coal, oil, gas and large-scale hydro (no nuclear), then we can have transportation that causes 90 percent less pollution and depletion than it does today.

Furthermore, electric vehicles are inherently more economical than current modes of transportation, if the accounting is done on a "level playing field" - which is far from the case now. The same obvious and hidden real cost factors that I outlined in the previous chapter also apply to gasoline-powered vehicles. In 1990 the Rocky Mountain Institute estimated that the real cost of oil from the Persian Gulf was $106 per barrel (42 gallons). If we take these true costs and problems into consideration (which of course we must), there can be no doubt that PV-powered electric transportation can be less costly, safer and more reliable.

My Electric Car

For several years I hoped that my old car would last until I could afford to buy an electric one. It had been suffering from terminal rust for a long time and was costing me about $1,200 annually for gasoline, oil, maintenance and repairs. In September 1994 my car suddenly smelled of something awful and the radiator spewed out gray foam: the engine oil was leaking into the cooling system.

The next day, looking through the Vineyard Gazette for a secondhand car, I found this ad: "Electric car. 645-...." I immediately called up, and the next day was off on a test drive. The car was an '81 Honda Civic, gutted, restored and retrofitted to be 100 percent electric. It looked and drove like a regular car, and with visions of powering the 17 batteries with solar PV panels placed on the roof of my garage, I instantly fell in love.

But the price was far more than a standard secondhand car. I had to seriously rein in my enthusiasm and do some calculations to compare the economics between a secondhand regular car and this electric car. My calculations revealed that this electric car would cost just about the same to own and operate as a gasoline car. The purchase price would be considerably higher than a good secondhand car, and thus the annual loan payments, but the costs for power, maintenance and repairs would be considerably less. I bought the car, with the condition offered by the real gentleman who sold it to me: if within three months I found that this car did not provide satisfactory transportation, I could return it for a full refund.

I got the car in early November and immediately loved it. The extra 500 pounds for the batteries made the car feel good and solid. Acceleration was just a bit slower than in a regular car, but regular driving was surprisingly normal.

The range per full charge was 50 to 60 miles, far enough to get anywhere I want to go on the Vineyard and home again. For that rare off-island trip not serviced by bus, train or plane, I would rent a car.

My plan was to cover the south roof of my garage with 144 square feet of PV panels (8 by 18 feet). At 15 kwh per square foot per year, this would generate about 2,100 kwh per year, and since the car required one kwh of electricity for every three miles of driving, this would generate more than enough power to drive my annual 6,000 miles. The PV would be interfaced with the power company, so on sunny days the excess PV power would feed into the grid, whereas on cloudy days the batteries would draw from the grid.

This particular electric car was a homemade conversion and lacked some important controls, such as an effective control for shutting off the electricity when the batteries were full. This was a real drawback because the longevity of batteries depends on how well they are charged: they need to be fully charged, but overcharging causes stress and therefore shortens their life span. It also wastes electricity. But the most serious drawback was the fact that when cold weather arrived, the range of the batteries was reduced by more than 50 percent. This problem could have been solved by enclosing the batteries with high-density foam insulation and providing low-watt electric heating within those battery compartments. But I could not afford to spend either the time or the money to improve the car. Since this electric car turned out not to provide reliable transportation in cold weather, I returned it, and instead bought a secondhand car with a regular internal combustion engine. My fervent hope is that electric cars will soon be mass-produced and therefore more affordable.

Thus my personal experience with the electric car was far less successful than my own various solar-dynamic, bio-benign experiments. Nonetheless, I believe more than ever that we have the technology, right now, that can provide us with better transportation than the gasoline or diesel engine, and at far less real cost. According to the latest update, batteries now exist that can be supercharged in three minutes and have a range of a couple of hundred miles. This is comparable to the time it takes to fill the gas tank and the distance you can drive before a refill. Gas stations could have this technology installed, and the PV panels could provide welcome shade and shelter. I believe that the electric vehicle, powered by PV and other renewable technologies, today stands ready to be the safest, least costly, least polluting, least wasteful, most reliable and most dependable form of transportation.

The other aspect of transportation that needs to be and can be vastly improved is public transportation, and in the section "A CALL TO ACTION" I present a proposal that would drastically reduce the use of private cars.

SOLID WASTE MANAGEMENT

When I first came to the Vineyard back in 1958 I summered in the town of Edgartown. It was customary then, there as well as in other communities, to throw all household waste materials together into trash barrels. There these wastes were transformed into a smelly mess called "garbage", which generated flies and attracted dogs, rats, skunks and racoons.

We took our garbage to the dump, which was a vast malodorous pile in an enormous deep pit. From time to time this garbage was intentionally set on fire to reduce the volume. Sometimes these fires raged unstoppable for days or even weeks, while the whole town was enveloped in foul smoke, and particles of greasy soot snowed down on our babies, gardens, cars, and the laundry on the line. This dump was heaven for flies and rats, seagulls and feral cats.

During heavy rains, the garbage pit would be deep in gunky, seething ooze, and as I looked at the tall municipal water tower standing close by, I wondered about the billions of molecules of harmful chemical substances that must be leaching into the groundwater, just a few feet below the base of the pit, getting pumped up into the water tower, and from there flowing into the town's water pipes and then into our bodies.

But I also loved the dump. In among the stale bread and broken bottles were treasures galore: lovely antique chairs and tables in need of just minor repairs, perfectly good clothes and curtains, lamps, books, toys, sinks and bathtubs. Once I found an old trunk filled with old lace, a worn but exquisite quilt, embroidered linens, love letters and family photographs, and even an opal necklace. And there were always endless amounts of building materials and kindling.

Then in the 1970s, the "dump" became the "sanitary landfill", as new regulations required that the garbage be covered daily with dirt in an effort to reduce odors, flies and rats. But the toxic molecules from dumped solvents and pesticides (there was DDT, 2,4,5T and all kinds of other stuff that has since been outlawed) were still leaching into the groundwater. Rescuing the treasures became much more difficult as seekers were chased away by a bulldozer operator determined to cover it all with dirt.

Those of us who were appalled by the waste and pollution worked for years to do more and more recycling, but often our carefully separated newsprint and glass, tin cans and aluminum were buried rather than shipped off to be recycled. We also, along with many other community groups around the country, campaigned for better management of toxic chemicals to prevent contamination of the groundwater, but this effort resulted in rather bizarre new regulations nationwide. On the Vineyard, "Hazardous Waste Collection Days" were set up on a Saturday morning twice a year, but because of the inconvenience and confusion about the location and the time, most of the hazardous wastes still ended up in the dump and thus in the drinking water.

The whole hazardous waste collection site was covered with plastic tarps, and staff and trained volunteers wore plastic moonsuits. Jars, boxes and cans, empty or containing remnants of toxic materials such as pesticides, herbicides and paint thinner, were placed, each surrounded with vermiculite, in 55-gallon drums. Paint was poured into 55-gallon drums. Then the drums were taken off-island, at $400 each. When I asked where the toxic load was destined, the community organizing team did not know, but the trucker thought most of it was trucked to Alabama for "treatment".

Numerous phone calls later I finally learned that this "solution" to the hazardous waste problem consisted of shipping our hazardous chemical wastes to incineration and burial facilities located in poor neighboorhoods in Alabama. These facilities caused horrendous pollution, and because of the shockingly high cancer rates that resulted from people breathing the air and drinking the water, the surrounding area became known as "America's Cancer Alley".

When this scandal was exposed, dumping was rerouted to Africa, Asia and islands in the Pacific. Here, because of the lack of regulations and enforcement, loads were often dumped, in exchange for minor but coveted dollars, without even informing the recipients of the degree and nature of the load's toxicity.

As news of garbage toxicity became better known, the NIMBY (Not In My Back Yard) syndrome spread, and it became increasingly difficult to find places to dump. Who can forget the images of garbage-filled barges being towed for months on end (some for 18 months) back and forth across the world's oceans in search of a place to dump their malodorous and toxic loads. Some dumped illegally in the ocean or on remote pristine beaches in the Caribbean and the Pacific, while others were forced to bring their loads back home.

Incinerators became the big thing in the 80s, some designed to generate electricity in the process. At first, many thought this was a great solution - voila! turn garbage into electricity. But about 25 percent of the original garbage remains as toxic ashes that must be buried in lined landfills that again and again have proved to leak into the groundwater, and the emissions from the incinerator smokestacks contain dioxin and other highly toxic substances. Furthermore, these incinerators destroy valuable resources that need to be recycled.

In spite of the dangers, expense and waste, the solid waste management committee on the Vineyard voted to make a contract with a new off-island incinerator and to ship to it as much trash as possible as soon as possible. Recycling, in the 80s about 5 percent, was perceived by most as never being able to go any higher and was therefore judged not to be worthy of serious consideration. The cost of garbage disposal went from around $25 per ton to over $100 per ton.

Because of continuing pressure from those of us who believed we must help stop the scandalous waste and pollution of precious resources, bins and trailers were gradually set up to hold more categories of recyclable wastes. It started with newspaper, glass and aluminum, and soon magazines and corrugated cardboard, one kind of plastic, and tin cans were added.

The cost of trash management was traditionally paid out of taxes, thus dumping trash had been "free". But soon a fee of $2.50 was levied on each barrel or bag of trash, whereas depositing recyclables was free. With this incentive the rate of recycling started to slowly rise. A few communities in the U.S. are now recycling over 50 percent, but the rate has remained far lower in most places, including the Vineyard. As long as individuals have to separate wastes into too many categories and to remove caps, rings and labels and flatten metal cans and plastic bottles, the recovery rate will remain low, and vast quantities of valuable resources will continue to be lost in landfills and incinerators.

The market values of recyclable materials go up and down, but have basically risen as more and more people demand goods with high recycled contents. Thus manufacturers want to buy more waste materials than they are currently able to get. The time is ripe for truly comprehensive recycling. I believe that we can reach 90 percent collection of recyclable materials (including the compostable and reusable portions), if we make it truly convenient for people.

Comprehensive Recycling

For years I have managed my household wastes in such a way that I recycle at least 90 percent of my wastes, contributing less than four barrels per year to the landfill or incinerator. This compares to 35 to 45 barrels per year for a comparably sized household doing little or no recycling. All compostables go into my compost bin, even meat and fat. There they are magically reduced to superb-quality compost that benefits the flower gardens.

Most of the plastics, all the metal and glass, glossy and colored paper and corrugated cardboard go to recycling, while the low-grade wastepaper gets compacted and used for winter heating (currently one of the best uses of low-grade paper). I stockpile those plastics that are not now accepted for recycling, because every molecule of plastic is recyclable again and again, and I therefore have faith that someday they will all be recycled. This scheme results in hassle-free, convenient, easy, odor-free waste management, and it takes no extra time. In fact, it saves me time because I need to go to the transfer station/dump only two to three times a year. And it saves about $100 a year in dump fees.

Here is a nice example of the result of the recycling we practice at Solviva. After the Solviva greenhouse was completed in 1984 we had a big celebration. About 130 people came for eight hours of general merriment. We had a five-course dinner, campfire, contra dancing, speeches and ceremonies. The next day we cleaned up, separating everything into four categories: returnable and recyclable bottles and cans, compostables (to enrich the gardens), burnables (to contribute to heating my home in winter), and reusables (washing all plastic cups, plates and utensils for reuse again and again over many years). There was only one object that could have been labeled as trash: a plastic fork that was partially burned when it lay close to the campfire. But this instead became the center pendant in a commemorative necklace I strung. If all this waste had been treated as trash, it would surely have filled at least six trash barrels, which at $2.50 per barrel would have cost $15.

I have spent thousands of hours over the last 15 years tracking the market value of recyclables. I have visited and corresponded with communities across the U.S. and around the world and have researched what equipment is available and what it costs to buy and operate. I have been calculating, planning and designing to find the best, most practical, most economical ways to manage solid wastes in a small community or a big city. In the section "A CALL TO ACTION" I present a proposal for solid waste management that would result in 90 percent recycling as well as substantial reduction in cost.

❦ ❦ ❦

PART FOUR

A CALL TO ACTION

PROPOSALS FOR JUMP-STARTING A BETTER FUTURE

❧ ❧ ❧

The Solviva designs, as well as many other designs developed around the world, have demonstrated that our needs for heating, cooling, food, electricity, transportation, wastewater and solid waste management can be fulfilled in ways that are far more reliable and secure, and are far less costly, polluting, wasteful and harmful than conventional methods. It is now time for us to be bold, to get up and do what needs to be done to protect ourselves and our planet, for now and for the future.

In this section I present proposals for the creation of "lighthouses" out in the "real world". If just one of each of these proposals were to be properly and fully installed, well managed, and thoroughly monitored and documented in order to prove the advantages, I believe that individuals, communities, cities and nations will then follow the lead. After all, why would any reasonable person prefer a system that is more costly, and causes more pollution and waste? We could thus see, within a decade or two, the 80 percent reduction in pollution and resource depletion that I outlined in the introduction.

Some people will consider these proposals hopelessly idealistic and unrealistic. In the "real world" perhaps they are, because unfortunately some people are mired in skepticism, pessimism, inertia, and an inexplicable terror of speaking up in favor of proposals they know in their hearts would probably turn out to be just as wonderful as they sound, but that cut against the grain of "business as usual". But I know one thing for certain: from a practical standpoint these proposals are all both achievable and cost-effective.

❧ ❧ ❧

Proposal No. 1

The Greening of the White House

A Proposal to Transform the White House into a Lighthouse to Guide the Whole World toward Sustainable Peace, Health and Plenty

Throughout the 19th century there were extensive greenhouses on both the east and the west ends of the White House. At that time, greenhouses attached to city buildings and mansions were prevalent all through the northern United States and Europe. These greenhouses provided the buildings with substantial amounts of solar heat, which on sunny days offered a respite from the constant stoking of wood and coal heating stoves. In addition, these greenhouses produced fresh flowers, fruits and vegetables during the cold season before the days of long-distance transportation.

By the end of the century, plans to expand the White House showed that the old greenhouses were to be replaced with new ones elegantly incorporated into the architectural design. But by early 1900, as central oil heating became available, all White House greenhouses were torn down and additions made to the east and west ends of the main building without greenhouses or solar heating.

I propose that the White House provide leadership in contemporary techniques for sustainable living by reinstating those greenhouses. The current White House complex has a long facade that faces due south. I am not proposing changes to the familiar look of the main central White House building, treasured as a national symbol. However, the long east and west wings on either side have no strong architectural or historical significance and are eminently suitable for solar-dynamic retrofitting. This proposal calls for restoring the historical greenhouses by retrofitting these wings with the best in solargreen design. This could reduce the heating fuel requirements of the entire White House complex by 70 to 80 percent. This can be done elegantly and without adding to the summer cooling load. Flowers, greens and vegetables can be grown as a model of nontoxic, high-yield indoor agriculture, also demonstrating how plants freshen the air with oxygen and negative ions and remove the carbon dioxide and infectious organisms emitted by people, as well as the toxic gases emitted by computers, carpets, paint and cleaning compounds.

In addition, this proposal calls for installing photovoltaic panels, interfaced with the grid, unobtrusively mounted on the roofs, to generate as much electricity as the White House consumes.

All wastewater would be cleaned through Biocarbon filters and dispersed through underground irrigation pipes to benefit the landscaping. This would reduce water consumption by some 15 percent and contribution to the city sewage system by 100 percent.

I propose that the Army Corps of Engineers or another military entity be employed to turn this dream into reality. With the training, skills, efficiency, and organization characteristic of the military, temporary encampment right on the spacious White House grounds, and superb planning and project management, a comprehensive solar-dynamic, bio-benign retrofit could be accomplished in a month or two. What better way could there be for the military to promote national security and peace around the world?

Imagine approaching the White House from the south, with the new elegant solargreen retrofits enhancing the east and west wings. Leaders from around the world meet with the President around the table under the dappled shade of fragrant jasmine, nasturtium and ferns, in a rooftop greenhouse above the Cabinet Room in the West Wing. School children tour the White House, meandering through the vibrant greenhouse gardens and rainforest landscapes in mid-winter, as a guide explains the benefits of these new designs and how they can be adapted to small and large buildings anywhere in the world. Imagine the inspiration and vision for a better future that will blossom within any person who encounters this reality.

PROPOSAL NO. 2

A SCHOOL THAT PROTECTS OUR ECONOMY, HEALTH, ENVIRONMENT AND RESOURCES TODAY AND TOMORROW

Public schools are community centers which are financed by taxes, and are therefore excellent places to apply innovative designs and technologies that offer better methods for heating, cooling and air purification, for food production, and for sewage and solid waste management. Not only does this offer communities a chance to save money and resources and prevent pollution, but it also offers opportunities for students, teachers, parents, and the community at large to learn important skills for good living. The time has come to add a few more "R's" to the curriculum: Recycling, Reuse, Resource Recovery and Restoration, Retrofitting, and perhaps most important, Responsibility, Respect and Reverence for Earth, life and coming generations.

Unfortunately, schools are being built and expanded in thousands of communities across the nation, including the Vineyard, with little or no regard for the needs of tomorrow, or even today. This is a tragedy. Over the last few years, five different schools on the Vineyard have been doubled or tripled in size. In the early stages of the planning for each one, I presented solar-dynamic, bio-benign proposals, which were repeatedly ignored and rejected by the architects who claimed that solar is neither affordable nor practical. This in spite of the fact that both my home and the Solviva greenhouse were available as eloquent proof.

Because of continued lobbying, some of the schools did get some very small attempts at solar, but my input was refused. As a result, these attempts are very costly and unsuccessful. Due to the architects' lack of knowledge and experience in comprehensive and efficient solar design, these token attempts actually consume more energy than they provide, thus perpetuating the impression that solar design is impractical and expensive. I am tempted here to express my frustration by writing pages about the extraordinarily expensive, impractical, wasteful, polluting, unattractive, unhealthy systems these architects installed. The schools now have ventilation, heating, cooling, lighting, and wastewater management systems that would never have been built if committee members had considered the real impact for both now and the future. Perhaps if architects were rewarded for cutting costs for both construction and operation, rather than earning a fixed percentage of the total cost of construction, they would have provided a more economical, practical and energy-efficient plan.

These five Vineyard school projects could have been built to consume 80 percent less energy, reducing depletion of oil by some 200,000 gallons per year, reducing annual CO_2 pollution of our atmosphere by 4 million pounds, and reducing the tax burden in our community by $200,000 per year.

Comprehensive solar-dynamic design could have been installed, complemented with multi-fueled, clean-burning, automatic furnaces, fueled primarily by wood chips from planted energy forests and deadwood from the adjacent State Forest, plus shredded low-grade wastepaper, with oil only as an occasional backup. Solar-dynamic design could have provided not only these savings, but also a strong sense of security for the whole community, stemming from the knowledge that these schools would never be threatened with closures due to scarcity and crippling cost of oil caused by future oil embargoes, wars or terrorism (certain to occur again and again). Also, the community would gain a strong sense of well-being by minimizing negative impact on economy, environment and resources, thereby protecting future generations.

My proposal for a good school design has south-, southeast- and southwest-facing walls and roofs that provide solar heating and cooling, with long-term solar heat stored in slab foundations, and ducts and fans distributing the heat as needed. Some of these walls double as greenhouses, with vegetation ranging from salad greens, vegetables, tomatoes and herbs to tropical flowers and vines. A brief period each week is enough to provide the opportunity for students to learn how to grow organic food and ornamental plants year-round. It also enriches the curriculum for art, sciences and vocational education. Students can gain experience in business and marketing by selling the produce, raising money for school trips and other "nonessential" expenditures. As a fringe benefit, the school population benefits from cleaner air, great places to study and hang out during breaks, and fresh organic salads every day, all of which improve the health and energy (and therefore the performance) of the entire school population.

Stale air from the classrooms is purified as it flows through these greenhouses, because plants have an extraordinary capacity for absorbing not only the CO_2 and infectious germs emitted by people, but also other air pollution, such as formaldehyde and other outgassing from carpeting, plastics and various cleaning compounds. Fresh air is also brought in from outside, warmed as it passes through the solar walls and roofs. These are capable of warming the air even on cloudy days.

Flush toilets and sinks drain into odor-free aerobic composting chambers. Here the solids decompose into superior compost, while the liquid, about 4,000 gallons per day for a 700-student school, drains through the compost chambers and into a pump chamber. Float valve-controlled pumps periodically pulse 200 gallons to a series of Brownfilters, housed in a small building. From there the odor-free effluent is pumped to perforated pipes laid in landscape-enhancing Greenfilter beds. This wastewater system costs 25 to 50 percent less than conventional septic systems, causes 90 percent less nitrogen pollution, and produces holly, Christmas trees, cedar posts, bamboo stakes, wood chips for heating, and valuable compost.

The food wastes and lowest-grade paper wastes turn into compost for the greenhouses and landscaping. The high-grade waste paper is recycled, as are bottles, cans and all plastics. All toxic wastes are kept separated and less than 10 percent of the solid wastes remain as trash, which can be safely landfilled on-island.

If a community such as the Vineyard, with a year-round population of about 14,000, had schools like this, taxpayers could save not only $200,000 in energy bills, but also many thousands on food that can be grown instead of bought, and many thousands on sewage and solid waste management, while at the same time improve health and education - and safeguard our future. This is not just a pie-in-the-sky dream, as many would think. This is practical and cost-effective today. It is of course best to build the solar-dynamic, bio-benign design right to start with, but it is also possible to retrofit almost any existing building, no matter how inefficiently designed and built.

Proposal No. 3

An Ecocomunity Recreation Center

This is a proposal for an Ecommunity Center for Martha's Vineyard, to be built adjacent to the high school, easily accessible by the entire community. The project is to be a model, to be followed by other cities and communities around the nation and the world. Heating, cooling, electricity, pool sanitation, and wastewater management will be of optimum solar-dynamic, bio-benign design, and both construction and maintenance will be at record-low cost. This Ecommunity Center will demonstrate the feasability of constructing buildings that cause 90 percent less depletion of energy resources, cause 90 percent less pollution of air and water, and improve public health, the economy, and national security.

The plan is to complete construction within 90 days of groundbreaking, at 40 percent less cost than normal. The major reason for the speedy construction and the reduction in cost is that the military will provide the key work crew. It seems reasonable to request that a small part of military training time be spent building for peace instead of preparing for war, helping to create municipal infrastructure, in community after community, that reduces the nation's dependence on foreign oil and other resources, thereby improving national security and the economy. Since the crew's time and support is already paid for by tax dollars, the cost to the communities would be minimal.

In addition, local builders and tradespeople will be employed, at 10 percent less than normal rates, and materials will be bought through local dealers at a 10 percent discount. Free family memberships will be given as compensation for these discounts. Construction will be further augmented with regular and frequent Amish-style barnraising events, culminating in fun-and-fundraising dinners.

A budget of $3 million is set for the cost of turnkey completion of the project and first year operating costs. Funds for the project will be generated from federal, state and municipal sources, local and national charitable organizations, and prepaid memberships.

The entire community is invited to participate in the planning process, including school children. A location is chosen where the planning process can evolve continuously over a period of two months. A basic plan is first presented, with models, architectural drawings and floor plans. Soon ideas from anyone wishing to contribute fill the walls. Classes from elementary schools and the high school, as well as various civic groups, come in to learn about the design concepts and add their own ideas. The pros and cons of all the various ideas are discussed, and those ideas that improve upon the basic plans are adapted and incorporated into the rapidly developing designs.

A proposal and petition is prepared by a group consisting of influential members of the community, outlining the plan and stating the many advantages of such a project. This is signed by most people in the community and endorsed by city, county and state politicians, as well as business, health and education professionals. This proposal is sent to the President, the Senate, the Congress, the Secretaries of State, Armed Forces, Health and Human Services, Environment, and Energy, requesting funding and participation by the military. Before the start of construction all materials for the entire project are assembled, catalogued and organized in the order they will be needed. On day one, June 21, a platoon of military personnel arrive and set up, on one of the large playing fields adjacent to the building site, their tents and other support structures, including a very large tent with tables and chairs for community dinners. On that same day a team of bulldozers and backhoes from several local construction companies dig the hole for the foundation and swimming pools. At 6 p.m. a couple of thousand people gather, at $10 a head, for the groundbreaking celebration and dinner, food provided at discount prices by the local markets, restaurants, and farms, prepared by the students and culinary arts department of the high school.

Over the next 30 days the military platoon work shoulder to shoulder with local builders and tradespeople with unprecedented efficiency, augmented by a steady stream of volunteers of all ages. A crew of well-coordinated supervisors keep everyone continually occupied. There is work even for children and people lacking construction skills, picking up and organizing nails, screws and wood scraps. Three nights a week there are community dinners, with celebration and progress reports, which raise both funds and spirits.

In one month the building and all its components have been completed, including the photovoltaic panels and paving-block parking lot. The progress has been recorded by video crews from near and far. CNN and other major networks have aired it around the nation and the world, along with requests for donations to a fund to help other communities start similar projects. Now, they are again on hand for the celebration of the completion of the first stage of construction and the emotional farewell to the military personnel.

During the next 60 days the final details are completed, including the interior rainforest landscaping and the salad garden, and local artists and children cover some of the walls with murals.

On September 21, 90 days after groundbreaking, the Ecommunity Center is open for business, on schedule and within budget, ready to greatly enhance the joy, health and education of everyone.

I now invite you to visit this center, visualizing it with me. Driving up to the parking lot, we see first of all that it looks more like a park, with shade trees between the rows of cars. Instead of asphalt it is paved with hollow blocks filled with topsoil, mosses and low grasses which absorb and digest the various pollutants that leak from parked cars, thereby protecting the groundwater. Along the south side of the parking lot is a bank of photovoltaic panels which produce as much electricity as the center consumes. Excess power is sent backward into the electric company grid when the sun is shining, and the center receives power from the electric company when there is no sun.

The building is attractive, like an immense barn. Every square foot of south-facing wall and roof is receptive to the power of the sun. Most of the south wall is super-insulated glass through which we see the beautiful indoor gardens. The south-facing roof looks much like the standard metal roofs of Idaho or Vermont. But here the dark brown metal roofing is topped with long-lasting, super-transparent plastic glazing, which transforms it into a comprehensive solar roof that not only heats and cools the building, but also heats the water for the pools and showers.

We enter through the east doors into the spacious entry hall, welcomed by the fragrant warmth and beauty of the sun streaming through a flowering rainforest landscape. Water trickles down pebble-filled troughs and moss-covered ledges, into a stream and a shallow pond surrounded by ferns and water plants. Children are draped over the railing of the wide bridge, watching fish dart among the rocks. Similar landscapes are integrated throughout the whole building, and all are part of the wastewater purification system.

In the cafe located in the southeast corner of the building you can choose a table in the sun or the shade. The menu, all organic, offers delicious meals (including hamburgers and hot dogs made with local organic beef and chicken), snacks, desserts, and fresh fruit and vegetable juices, as well as salads grown in the greenhouse that stretches along the south wall of the building. You can also buy food to take out.

Sign-in is by the office that opens to the entry hall, and in here are several monitors on which the entire building, including the pool areas, are displayed. This minimizes the number of lifeguards that are required. Close to the office is a well-equipped, supervised playroom for children whose parents are swimming or exercising.

In the northeast corner of the building are the dressing rooms. These contain the usual showers and lockers, but the antiseptic look of the tile walls is softened by more rainforest landscaping along the windows. The path to the pool area leads through a fern-lined corridor with motion-activated warm showers. Thus everyone is cleansed before entering the pool area.

The pool room is huge, warm and sunny. Speakers carry sounds that vary from wind, waves and songbirds, to Mozart or rock and roll. Three different pool areas are separated by walls of rainforest plants that not only provide beauty and air purification but also tone down the happy but loud voices of the swimmers.

The two smaller pools, 20 by 30 feet, are located along the south side greenhouse. One of these pools slopes like a beach into shallow water. Here toddlers, all wearing float vests, cavort in the nice warm water, dangle and plop down from grab bars, slither down little slides, giggle under fountains, and tumble on a variety of soft plastic play equipment on the deck around the pool. Some of the parents are in the water with their children, while others keep a close watch from lounge chairs around the pool. The floor, here as in the whole building, is laid wall-to-wall with firm foam mats to prevent injuries. Because this pool gets more contaminated due to diapers, the water has its own Biocarbon purification system to avoid burdening the main system.

The second pool, also warm, slopes to a depth of 5 feet and offers wheelchair access. Here older people, and others who would rather not be among the rambunctious crowds in the large pool, are floating or swimming laps in peace and quiet. Handicapped people are enjoying the freedom and exercise afforded by floating in water. People are basking in the sun in lounge chairs, surrounded by banks of flowering plants.

The third pool is Olympic size, with a movable partition to enable multiple uses. On this day the largest part of the pool is occupied by teenagers training for a swim competition. In the smaller section of the big pool another team is perfecting their water ballet. Yet another team is practicing dives from three different levels of boards. The tallest board is up on the mezzanine that surrounds the large pool. At the other end of the mezzanine is a long winding slide that leads into the pool.

Main access to the mezzanine is via a ramp that winds along the interior walls of the building. This ramp continues up to the smaller third floor and provides wheelchair access to the entire building. The ramp also doubles as a running track. People can clock in miles by running up, around and down this track.

The mezzanine floor is open over the large pool, and the sides are angled to create seating for spectators. Also on the mezzanine floor are two spacious rooms. One is filled with a great variety of gym equipment, most of which is connected to generators. Thus people are producing electricity while they pump away on weight or rowing machines, Stairmasters or bicycles. Users insert their membership cards into each machine, which records the amounts of watts generated and earns credits toward facility fees. On an average day this room can generate 100 kilowatt-hours, over 36,000 kwh (36 megawatts) per year.

Another room is for gymnastics, aerobics, dance, yoga, tai chi and other fitness classes.

The ramp continues to a smaller third floor which has Jacuzzis and saunas, as well as several small rooms for different forms of therapeutic massage and treatment.

Maintenance equipment is located on the first floor along the north wall. Here pumps drive the wastewater from toilets and showers through Biocarbon filters which render the wastewater clean and odor-free. The pool water is filtered and then purified by ultraviolet radiation and ozonation, rather than toxic chemicals. The purified pool water is heated in pipes laid within the solar roof when the sun is shining and at other times through pipes within the furnace. This is a clean-burning furnace, fueled by chipped wood and wastepaper fed by thermostatically controlled augers, which provides the backup heat for the whole center.

This Ecommunity Center offers true recreation for the whole community year-round. Families, children, teenagers, elderly, single people and married come here to socialize and have fun, to relax and get fit, and to eat good food. Because its warmth and cleanliness are maintained without burning the many thousands of gallons of oil normally required, without using chlorine or other toxic chemicals, and without causing water pollution, it is also a place of great inspiration where everyone can learn how to live in wonderful ways that save money and protect Earth's environment and resources.

❦ ❦ ❦

Proposal No. 4

A Solviva Restaurant and Business Center

This is a proposal for a business center that offers delicious and wholesome food at reasonable prices, as well as entertainment, education and culture, for all ages, in a delightful atmosphere and beautiful setting, good food for body, mind and spirit. I invite you to come along on a guided tour of this Solviva Restaurant and Sanctuary, yet another dream of what is possible to achieve. It occupies an area 200 by 130 feet, less than two thirds of an acre, which includes a restaurant, gallery, store, farm, offices, and parking for 40 cars, two delivery trucks, plus bicycles.

Imagine a very cold, sunny day. We approach along the south fence, topped with photovoltaic panels which generate as much electricity as is required by the whole complex. We turn right along the west fence, and as we enter the gate, a rooster crows. The landscape of gardens, trees, benches and paths is now under a blanket of snow. The main building is at the north end of the property, toward the left, but we first follow the path to the right, toward the sound of the rooster. We cross a little bridge over a pond and brook. Fish swim in the water, which is kept from freezing because it is agitated by a small windmill and warmed by a solar panel.

Here at the south end of the property is a mini-farm, with a couple of happy ewes and their newborn lambs, a burro, and a flock of hens and the rooster, all munching grain, hay and fresh greens. On the west side of this mini-farm is a small energy-self-sufficient greenhouse/barn. A waterwall separates the animals from the plant section, where a continuous supply of fresh sprouts for the animals, as well as salad greens and herbs for the restaurant are produced. The farm is supervised, so while parents linger in the restaurant, children can play with the animals or on the adjacent jungle gym.

We head back past the snow-covered gardens, then across a wide deck covered with a clear roof and an arbor draped with bare grapevines.

As we step into the main building through the double airlock entryway, we are embraced by the comforting solar warmth and the enchanting fragrance of fresh baked bread, flowers, and living earth. To our left the sun is streaming in along the whole south wall through the vibrant greens and herbs that fill the raised beds and hanging growtubes. A profusion of flowers and vines cascades from ledges.

Overlooking the indoor garden is the dining room with 24 tables. Here oiled wood and yellow and warm earthtones provide a comfortable and cozy atmosphere. The walls provide space for art exhibits. Near the entryway is a massive heating stove built into the brick wall. On days when there is no sun to keep the place warm, this stove does the job by heating up the brick wall and the whole building. The fuel is wastepaper briquettes and wood.

Right in front of the entryway is a reception desk and checkout counter. To the right of this is the store. The first wall of the store, right by the entryway, is dedicated to the display of information about solar-dynamic, bio-benign design, technology and methods. It explains how this Solviva building works, the heating, cooling, solar electricity, wastewater purification, nontoxic pest management. It shows designs of homes before and after Solviva retrofitting, and Solviva solargreen home designs of different sizes, as well as community centers, schools, and businesses. There is information about how much they cost to build, how much money and resources they save, and how much less pollution they cause compared with standard buildings. Here people can begin to dream and plan how they can lower their cost of living and improve their quality of life, and how they can learn to live in ways that have a positive instead of negative impact on Earth and the future.

The store offers a select choice of clean/green light/right products, such as energy-efficient light bulbs, recycled paper products, clothes and dry goods made from organically grown fibers, nontoxic body care and household products. Catalogues offer the best energy-efficient refrigerators and other appliances, as well as solar electric panels and equipment. There are books and magazines, cards and calendars on related subjects. This store also offers handcrafted goods by local artisans. In one corner is a loom, and a woman is weaving a shawl of soft angora wool yarn. Beautiful hand-dyed skeins of the same yarn, as well as sweaters, hats, blankets and shawls fill shelves close by.

A nearby counter opens to the kitchen, which offers takeout food: dinners, soups, quiche, breads, sandwiches, salads, herbs, fresh fruit and vegetable juices, jams, jellies, sauces, dressings and condiments, and luscious desserts. All foods are grown and made from ingredients that are free of any pesticides or other harmful substances. Good food, good for you. Some of the food is grown on the premises, and most of the rest, including meat, comes from organic farms near by: bioregionalism in action.

In one corner of the dining room is a raised stage area and a piano. Several evenings each week there are various cultural events, such as a string quartet, a jazz combo, a singer and guitar, poetry or story telling, a play, a movie, or a slide presentation. Sometimes the center tables are cleared away to make room for dancing. On weekend afternoons there may be a puppet show or play for children. And every couple of weeks there is an opening event for the art gallery.

In summer the gardens are bursting with a profusion of flowers, fruits, herbs and vegetables. Private seating areas are tucked away here and there among the shrubbery and flower beds. The large deck along the south wall of the building holds another 24 tables and a juice and salad bar. It is cool under a canopy of flowering and fruiting vines and protected from rain by a clear roof.

The second floor of the building contains the offices of several health professionals. On the third floor are apartments for the caretakers of the farm and the indoor and outdoor gardens.

All wastewater, including toilets, drains into Biocarbon wastewater purification filters. The cleaned water circulates through the canals and pond in the garden and is used for irrigating the ornamental landscaping. The parking area causes no groundwater pollution because it is paved with hollow blocks, topsoil and grass.

All solid wastes are recycled. Some of the food wastes are fed to the animals, the rest are composted and fed to the ornamental landscaping. The paper wastes are shredded and briquetted and used as fuel in the massive heating stove, with comprehensive smoke filtration to prevent harmful emissions. The glass is crushed and stored. The corrugated paper, metals and plastics are separated, compacted, baled and stored. Periodically they are trucked off to the highest bidder.

The entire Solviva Sanctuary complex causes no discharge of toxins or nutrients. It serves as a model to demonstrate how any existing building or new construction could be equally nonpolluting.

PROPOSAL NO. 5

SOLID WASTER MANAGEMENT THAT RESULTS IN 90 PERCENT RECYLING

Municipal solid wastes (MSW) consist of some 20 different categories: food wastes, leaf/yard wastes, clear glass, colored glass, five to six different categories of plastics, tin cans, aluminum, newspaper, magazines, cardboard and corrugated paper, plus large and small household items, appliances, construction wastes. At least 90 percent of these wastes are either compostables, recyclable or reusable, less than 10 percent is neither.

Most recycling systems are excessively complicated, requiring households and other waste generators to separate their recyclables into too many different categories, but in this system only four categories are required:

1. "RECYCLABLES and SMALL RE-USABLES" (about 50 percent of the waste stream): all clean paper, plastics (every molecule of plastic is recyclable again and again), glass and metal, plus books, clothes, toys, small household items and tools and so forth, all commingled in one container. Cans and jars need to be washed, but need not be crushed as this gets done at the separation center. There is no need to remove paper from every single tin can because it is not required by the tin reclaiming company.

2. "COMPOSTABLES" (about 40 percent of the waste stream): compostable wastes, such as all food wastes (including bones and fat), dirty paper (including paper napkins, plates and cups), diapers, and leaf and yard wastes, small compostable construction wastes (including wood and Sheetrock) all in one container.

3. "TRASH" (5 to 10 percent of the waste stream): whatever is truly unrecyclable, uncompostable or unreusable. This would include items such as the container of rotten food from the back of the fridge that you don't want to deal with, the mixed-material containers consisting of unseparable plastic, metal and paper (these will eventually be outlawed), dirty plastic, broken china.

4. "HAZARDOUS" (less than 1 percent of the waste stream): batteries, pesticides, paints, thinners, smoke detectors, flea collars and so forth. It would be against the law, enforced by stiff fines, to put any toxic wastes into any of the other waste categories.

In addition, there are the large re-usable items not normally counted as a catagory or percentage of the household waste stream, such as furniture, appliances, bath tubs, mattresses, rugs, construction materials, electronics, and so forth.

Bags or cans of categories 1, 2, 3 and 4 are put out at the curb and are picked up in a compartmented truck. The contents are checked and weighed on a computerized scale, and each household or business is billed accordingly. A small fee is charged for categories 1, 2 and 4, a much larger fee is charged for No.3. A reminder note is left in case of inaccurate separation. There are various attractive and fun economic incentives for good separation, fines only for serious infractions. Such a collection system can be put in place even in urban high-rise buildings.

Those who do not have pickup service take their wastes to one of several drop-off centers, or to the main Waste Management Center, where a similar procedure of checking, weighing and charging is in place.

At the main Waste Management Center are several large buildings, warm in winter and cool in summer because of the solar-dynamic design. One is for the separation, processing and storing of the No.1 category, the RECYCLABLES and SMALL RE-USABLES. At one end of this building all these clean commingled recyclable wastes and reusables are unloaded: the newspapers, magazines and junk mail, the plastic and cardboard milk containers, the cereal and pasta boxes, all categories of glass, plastic and metal, cans, bottles, jars, boxes, bags, foil and film, books, toys, tools, clothes, small appliances. It all goes together up a conveyor belt to the second-floor level.

Up on the second floor it is pleasant and clean. Sun pours in through the south window wall, flanked with boxes of flowering plants. The air is fresh and dust-free because it recirculates through Biocarbon filters.

A manager oversees and instructs a crew of people, more in summer, fewer in winter, who work here at good wages. Some work full-time, while others are part-time. Thus mothers and high school students, artists and writers, elderly and handicapped can sign up to fit their own schedules and abilities, gaining an opportunity to earn money and work constructively and collaboratively with others.

They work at different stations along the branching conveyor belts that advance as needed (some of the stations are wheelchair-accessible), and each person is responsible for pulling out one of 15 to 20 different categories. The separated wastes fall off the ends of their respective branches of conveyor belts and into chutes that send them down below to waiting compactors, shredders, crushers or carts. From there they go, in bales and hoppers, by conveyors or handtrucks, to protected storage to await shipping to various industries that buy the materials and remake them into new products. The re-usables go straight to the Community Re-use Building.

A second building is for the category No.2, COMPOSTABLES. Here a small well-trained crew converts the compostable solid wastes, plus septic tank pump-outs, into valuable compost.

Trucks and Dumpsters holding the food and dirty paper, diapers, leaves and yard wastes, plus trucks loaded with the lowest grade construction wastes such as woodscraps and bits of Sheetrock (nails are okay), enter the building and dump down onto a paved floor. The loads are bulldozed into mixed piles and are covered with compost, which immediately eliminates any odors.

The bulldozer pushes a well-balanced recipe of the mixed compostables toward a conveyor belt which shunts the mix into a big tub grinder. The mix comes out the lower end in pieces no bigger than two inches, and is then conveyed up into one of several composting drums. Standard 10-cubic-yard cement mixer drums (used ones with a few holes are fine) would be ideal and economical, mounted in series for simultaneous power turning. About 8 cubic yards of the ground-up compostables goes into the composting drum, which already contains about 1 cubic yard of live active compost, like yeast, left in the drum from the previous batch.

Incoming septic tank pump-out trucks empty their loads into closed tanks equipped with Biocarbon filter odor control, and, as needed, this septage, including the sludge, is pumped into the composting drum with the ground-up solid wastes, to add beneficial nutrients and moisture, which speed up decomposition. I know from experience that about 70 percent moisture is ideal for earthworms and other aerobic composting organisms. The whole batch spends four to five days in the chamber, and with periodic turning the temperature rapidly rises. It can easily reach 170 degrees F, but is kept below 100 degrees, ideal for rapid decomposition, but not so hot that it kills earthworms and other decomposers.

Then this coarse compost, free of unpleasant odors, is poured out of the drums and laid out in long windrows about 4 feet high and 6 feet wide. The first five rows are laid within the building, which is equipped with Biocarbon odor-control air filters to guarantee odor-free operation. Each row within the building, as well as the many rows outside, are turned over weekly. All the rows are thus advancing about 6 feet weekly, toward the final processing point. Over a period of two to three months, with the help of resident earthworms and beneficial microorganisms, this becomes a soft, fine-textured compost, with the fragrance of rich soil.

The compost is then triple-screened and tested for nutrients, toxins and pathogens. Compost that is to be used for vegetable production is placed on a conveyor belt that goes through a solar-heated tunnel which reaches pasteurizing temperatures. Thus any remaining harmful pathogens are destroyed. The compost is sold in bags or by the truckload, all tested and certified for nutrients, pathogens and toxins.

The coarse material that does not go through the screening process is brought back to the beginning of the cycle, rich in composting organisms, and is used as cover material to eliminate odors of incoming wastes, and thus goes around the cycle again for further breakdown. There will be a certain amount of noncompostable materials, such as plastic from diapers. The plastics will have been shredded in the grinder, but do not decompose. They will be screened out and recycled as lowest-grade plastic, ground up and melted, and made into all kinds of consumer goods, including plastic lumber.

The community of Martha's Vineyard, with a year-round population of about 14,000, swelling to 90,000 in summer, could produce around 8,000 tons of compost per year, which would be highly beneficial for Vineyard gardens and landscapes. Sold at or below current market value of good compost, it could have a market value of about $1 million. This, together with dumping fees, would more than cover the cost of capitalization and operation. This money would recirculate within our community instead of being bled off-island to pay for incineration, septage and sludge treatment, and for imported humus and compost. The income from the sale of compost and the fees for dumping the organic wastes and the septage (which could both be lower than current rates) could make such an operation a profitable business.

There are Dumpsters or trailers for dumping category No.3, the TRASH. This remainder will amount to less than 10 percent of the waste stream, and because it will contain no toxic materials, it can be safely landfilled on the island, thus eliminating the need for trucking off-island.

Category No.4, HAZARDOUS WASTES, goes into a secure, fireproof concrete and steel building with leakproof floor. The building contains rows of steel shelves, bins and drums for storing the sorted and indexed toxic wastes. Here leftover paints, screened and mixed into pleasing colors, screened thinners, and anything else useful, can be picked up at low cost. All toxic materials are closely monitored and safely stored. Some gets shipped off, but only if a dependable, safe, nonpolluting processing facility can be found.

The larger re-usables are kept well organized in yet another area for community exchange. Here is another large building with shelves, tables and bins for sorting and organizing clothes, blankets, curtains, furniture, books, appliances, electronics, tools and gadgets, sewing and hobby stuff. Children's books, toys, furniture and clothes are in an enclosed area in one corner of the building, which also opens onto an enclosed outdoor area. Here children can safely play while parents browse.

An area adjacent to this building is available for weekend flea markets. Unsold items can be left at the Re-use Center.

Also adjacent is an expansive outdoor area, some of it under roof, for all kinds of re-usable construction materials, well organized. Unseparated loads are weighed and charged fees sufficient to cover the salaries of the staff required to do the separation and organizing. No charge is levied on properly separated and placed stuff. Categories include: re-usable lumber, kindling wood, particle board and plywood, Sheetrock, insulation, doors and windows, stoves, refrigerators, bathtubs and sinks, and so forth. There is no more requirement for time-consuming and wasteful breakup of lumber into pieces less than 4 feet long, no more need to pay $100 per ton or more to ship and tip at the off-island incinerator, no more wasting thousands of dollars' worth of re-usable building materials, no more expensive trucking of re-usable appliances, bathtubs and sinks off-island for disposal.

This same system of collection and processing can be set up in cities and communities across the nation and would result in about 90 percent recycling, preventing needless pollution of air and water, destruction and depletion of trees, oil, aluminum, tin, and so forth. It would also create many jobs and improve the economy. And it would even make life more convenient and fun.

❦ ❦ ❦

PROPOSAL NO. 6

PUBLIC TRANSPORTATION

When I was a child growing up in Sweden I could get anywhere I wanted to go via buses, trams, trolleys or trains. They were frequent, well coordinated and on-time, and because they were mostly electric, they were clean and quiet. Public transportation was much used because very few people had cars.

In Curitiba, Brazil, the city's dynamic mayor, Jamie Lerner, decided something had to be done about the miserable congestion and pollution caused by thousands of cars and buses. With skillful planning and amazingly fast implementation, he created what is probably the most efficient urban transportation system in the world. One of the most important keys to its success is the loading system. People pay at automatic stiles as they enter loading shelters at every bus stop, much like subways. Thus, the buses progress very rapidly. The system is affordable and much used, and has drastically reduced traffic congestion and pollution. This system is described in detail in Bill McKibben's excellent book, HOPE, Human and Wild.

The following is a proposal for a public transportation system for Martha's Vineyard, adaptable to any other community. This system has been developed through many years of research, as well as contemplating what kind of public transportation system would make me choose not to use my car.

The system consists of three main components:

1. Run electric vans every 5 to 20 minutes, or as frequently as demand dictates, in several connecting loops that cover most of the island. Establish battery recharge terminals, powered primarily by banks of photovoltaic panels, at the various island landfill sites, which also happen to emit methane gas that can be burned to supplement the solar power (this would also help prevent global warming).

2. Establish turnoff-pickup spots approximately every quarter to half mile, each with a small shelter with maps and a change machine. Charge $1, less for the young and the old.

3. As an auxiliary to the vehicles that are dedicated specifically for public transportation, establish a system whereby anyone can apply to become a permitted pickup vehicle. Thus, Jane Doe can have her vehicle approved, registered and clearly marked as being available to pick up riders. You might call it a licensed hitchhiking system. Jane Doe sees a person signaling at a turnoff, and, if she wants to, swings in to pick up. She gets a dollar per ride, more if she is willing to go out of her way.

If such a transporation system were in place, I would often choose not to use my own car, but would instead prefer to walk down the neighborhood driveway and then to the nearest pick-up-turnoff. This system also calls for narrow walking paths meandering through the woods and fields along the roadside. Sometimes I would even prefer to walk a mile or more before stopping for pickup.

I believe that such a public transportation system would be cost-effective for users. It would be convenient because there would be many pickup locations and frequent pickups. It would significantly reduce downtown traffic congestion, and I wouldn't be surprised if closer scrutiny would reveal that it could also be a break-even, or better, regional municipal service, perhaps even a profitable private business.

❀ ❀ ❀

A TALE OF TWO CITIES

GRAYBERG or GREENDALE: Where Would You Rather Live?

At the risk of being repetitive, I now want to present a comparison between two hypothetical cities, summarizing the difference between the conventional "hard path" ways of living and the solar-dynamic, bio-benign "soft path" ways.

The first city is Grayberg, and you will recognize many of its methods of providing heating and cooling, wastewater and solid waste management, food, electricity, and transportation, because these are indeed the methods that prevail today in most towns or cities in the United States and the rest of the industrialized world.

The second city is Greendale, and although the residents there have the same needs as the residents in Grayberg or any other city, the methods by which these needs are satisfied and the resulting quality of life and cost of living are very different.

The cities have the same climate conditions, and both have populations of around 2 million. This tale is set just a few years hence, let's say the year 2005, because I believe the transformation could be accomplished in one city within seven years.

First let us visit Grayberg.

This city is located near the ocean, around a lagoon, in a valley surrounded by beautiful mountains. However, the mountains are rarely visible because of smog, much like Los Angeles. The smog is created by emissions from the tailpipes of cars and the chimneys of homes, schools and industry.

The city is heated primarily by oil, over one billion gallons of oil annually (an average of 500 gallons per person per year, which includes a share of public buildings and businesses), at a price which is about the same as it was in 1996. (For simplification let me say $1 a gallon, less than bottled water!) This totals over $1 billion per year. Burning this oil causes almost 10 million tons of CO_2 emissions per year. (Almost 20 pounds of CO_2 is produced by burning one gallon of oil, even with clean-burning furnaces and engines.) Many have installed woodstoves in order to reduce heating bills, but, due to serious air pollution, their use is banned many days each winter. The heating costs for the school district alone amount to over $30 million per year. Little is left in the school budget for "luxuries" such as arts, trips and new equipment.

In the bitter cold winter of 1999 Grayberg was severely impacted for several months because of yet another oil embargo. As in 1973-74 and 1979, prices skyrocketed, gas pumps were empty and people froze. This oil crisis resulted in the United States again going to war in the Middle East to secure continuing and ever-increasing access to the oil supplies. The oil came back, but this war was far more devastating than the one in 1991.

The oil crisis and war resulted in lifting the preexisting ban against oil drilling off the coast and in the wilderness beyond the mountains. Oil spills ruined the fishing industry, and oil washed up on the beaches, severely impacting the tourist industry. The wilderness and mountains, which had previously provided a haven for fishing, skiing and hiking, were devastated by trucking roads, pipelines and drilling towers. In an attempt to improve the economics of the town, the ban on clear-cutting forests was lifted. Driving along the road, the forest still looks grand, but just behind this facade stretches mile after many mile of denuded hills and valleys, deeply scarred by erosion ruts, debris-filled streams running yellow, red and brown among them. The few forest areas that are still intact suffer from air pollution due to acid rain.

Several million pounds of nitrogen annually leach into the groundwater from the city's sewage treatment facilities and the thousands of on-site underground septic systems around the outskirts of the city. The nitrate level in the drinking water has been rising steadily to the point that it threatens public health because it reduces the blood's oxygen exchange capacity, and many pay dearly for bottled water imported from distant springs.

The groundwater is also being contaminated with oil leaking from underground tanks, as well as hundreds of different chemical effluents from the various local industries.

The lagoon, previously rich in fish, clams, scallops and oysters, has been closed to fishing and swimming due to contamination. This contamination is caused primarily by the nitrogen leaching from sewage and septage systems via the groundwater, which in turn causes massive algae growth that suffocates the aquatic flora and fauna and putrifies and oozes up to the surface, periodically causing vile odors that envelop much of the city.

These algae infestations were first blamed on ducks, geese and other water birds and runoff from roads, lawns and farms, but now it is understood that the primary cause is the effluent from the conventional sewage and septic systems.

Not only are these polluting, expensive conventional wastewater management systems allowed, but they are legally required. Requests for permission to install various alternative technologies have been delayed and obstructed in numerous ways, while millions of dollars have been spent on team after team of engineers who have proposed ever more complex conventional technologies. Grayberg recently spent over $500 million for rebuilding the old sewage treatment facility to reduce nitrogen contamination and odors, and also to extend the sewer pipes to include most of the areas that previously had on-site septic systems.

Grayberg has three public recreation centers with Olympic-size swimming pools, built before the city fell on hard times. They are heated entirely with oil. During the oil embargoes and war they were closed for an extended period, but are now open again, with an annual heating bill of over $400,000. The pool water is sanitized with the usual chlorine, bromine and other toxic chemicals. This, along with the city's air and water pollution, is considered a major reason for the ever-increasing rate of allergies and asthma.

Supermarkets carry food from far distant places, most of it heavily laden with residues of many different toxic chemicals. Some food is produced locally in summer, none in the winter.

Public transportation, consisting of smelly, noisy diesel buses, is inconvenient, slow and unreliable. The few bike paths do not lead to where people want to go. Consequently, cars are needed to get to work, schools, shopping and recreation. Pollution and traffic are a nightmare.

The city does some recycling, but since methods of collection and processing are inconvenient, it has never amounted to more than 20 percent. Two million tons a year of solid wastes go to the incinerator, at $100 per ton, or $200 million a year. Citizens are alarmed that the prevailing winds bring back to the city emissions from the smokestacks of this incinerator, containing dioxin and other toxins, which are then breathed in by the people and stored in their bodies. On the advice of health officials, many mothers are sadly opting to feed their babies formula instead of their own contaminated breast milk. Of course, since the city drinking water contains high levels of nitrate, the formula has to be made with bottled water.

The city's electricity is generated partly by an oil-powered plant in the outskirts and partly by a nuclear power plant located across the mountains. This nuclear plant is located close to an earthquake fault, and radioactive wastes have been leaking into groundwater that is flowing toward the city's water supply.

The leadership of Grayberg is strictly opposed to any of the so-called alternative solutions. Any proposals that contain words such as solar, organic, bio-benign, harmonious, sustainable, renewable or alternative are rejected without even being seriously considered. Most people believe what conventional engineers and architects are stating, that such systems are too costly, unreliable and inconvenient. The few people who make proposals along those lines are labeled "impractical, unrealistic idealists". Whether it has to do with heating and cooling, wastewater or solid waste management, or transportation, when proposals are considered, the leaders choose costly, polluting, resource-consuming, business-as-usual methods, because, as they say, "we know it works".

Not surprisingly, Grayberg, once rated among the most beautiful and livable cities in the United States, is now low on the list. The quality of life, water and air is among the worst, while the cost of living, taxes, and health problems are among the highest.

This tale about Grayberg is all too familiar as it in truth describes the conditions in so many cities and towns in the United States. Most people do not know any other ways to live and are thus resigned to an ever-increasing cost of living and a gradual reduction in the quality of life.

Now let us visit the other city in our tale, Greendale. This city has the same climate as Grayberg and also about 2 million inhabitants. It too is located close to the ocean around a lagoon and is surrounded by mountains. But there the similarities end. The first noticeable difference is the clean air. The breathtaking view of distant snow-covered peaks, unobscured by smog, can be enjoyed even from the city center.

One reason for the clean air is the excellent public transportation system. Bright yellow buses shuttle back and forth every few minutes, topped with flags color-coded by route. All buses are electric, powered by batteries. At the end of each route are banks of solar photovoltaic panels that recharge the batteries. Other generators, powered by methane from the local wastewater treatment facility and wood chips from the construction industry and wastewater-fed energy forest, provide backup charging power. As the buses come in, their spent batteries are exchanged for fully charged ones, which takes only a few minutes.

Photovoltaic panels are manufactured in one of Greendale's many thriving factory complexes at the outskirts of the city. Batteries are manufactured at another factory, and, when finally exhausted, they are collected, disassembled and remanufactured. As with all other manufacturing in the city, any reusable or highly toxic chemicals are first removed from the wastewater, and then the wastewater is purified through Biocarbon filters and used for irrigating the city's extensive parks and landscaping.

Because of the excellent public transportation system, private car use is much reduced, and most cars and trucks are electric. Municipal parking lots are roofed over with banks of photovoltaic panels, and people put money in meters to recharge their car batteries while they are at work. The meters work even when the sun is not shining, because electricity is assured by backup sources.

Another reason for the clean air is that about 80 percent of the heating and cooling of the houses and other buildings is provided by the sun. The leadership in the city of Greendale found out about truly comprehensive solar-dynamic design in 1998 and proceeded to apply it to a new school addition. It quickly proved to be effective, reliable and cost-saving, and soon the city went all out to retrofit every school and other municipal building. This was found to be cost-effective immediately because the resulting savings were greater than the cost of financing the changes. It did not take long for the population as a whole to follow suit, and soon most of Greendale's buildings, old and new, were retrofitted to be heated and cooled primarily by solar power.

Backup heat is provided by clean-burning stoves and furnaces fueled with wood and low-grade wastepaper, and because the city is not producing the normal amounts of pollution caused by burning millions of gallons of fuel oil and gasoline, the air quality is not threatened by these emissions. Most of the wood fuel is wood chips, which are stored in hoppers and loaded into stoves and furnaces by thermostatically controlled augers. The wood chips are produced from rapidly growing shrub willows in energy forests, which are fertilized and irrigated by the effluent from the nearby wastewater treatment plant. Thus, the nutrients in the wastewater is utilized for making fuel, which purifies the wastewater in a most thorough and cost-effective way, protecting the economy, environment, fishing industry and public health.

Less than half of the city is serviced by the preexisting centralized sewage treatment plant. This plant was fully upgraded with Biocarbon filters at a cost of about 80 percent less than conventional technologies. And it costs about 90 percent less to operate, partly because the expenses are offset by the income from the resulting popular compost product, Greendale Black Gold, as well as from energy forest wood chip fuel which provides the backup heat for the city and methane for electricity production.

The people of Greendale avoided the enormous expense of expanding sewage pipelines, because on-site septic systems were instead upgraded with individual Biocarbon filter systems. By choosing these systems for upgrading wastewater management systems, instead of going with the conventional sewage, septage and septic systems that were chosen by Grayberg, the citizens of Greendale have saved some $200 million, and the groundwater and the lagoon are kept pristine.

A few years ago Greendale opted not to buy into a long-term contract with a planned new $5 billion nuclear power plant across the mountains. Instead the city invested in "negawatt" energy conservation, such as leasing out super-efficient appliances and light bulbs. This resulted in a 60 percent reduction in electricity consumption, thereby eliminating the need for increased generation.

Greendale also invited manufacturers of photovoltaic panels to set up operations in an abandoned pesticide factory. Calculations made it clear that it would be less costly in the long run, and far safer, to provide electricity by solar power instead of nuclear.

The Greendale electric company scoped out thousands of small sites within the city limits rooftops, walls, fences and embankments with good solar exposure - and installed PV panels. Far more electricity is produced than is needed when the sun is shining, and the excess solar electricity is used to generate waterpower: pumping water into towers and releasing it through generators when there is no sun.

Backup electricity is provided by the preexisting oil-powered plant, which was retrofitted to also burn wood chips and methane. This has proven to be cleaner, more reliable and less costly, and consumes 90 percent less oil.

To everyone's great relief, the plan to build a nuclear power plant across the mountain was canceled when Greendale, along with other cities, refused to join the contract.

Greendale, like Grayberg, also has recreational facilities with Olympic swimming pools. In fact, after the three preexisting facilities were retrofitted with solar-dynamic, bio-benign design, they became so popular and were so economical to operate, that several more were built. These facilities have not only one large pool each, but also two smaller, warmer pools, one for babies and toddlers and another for the elderly and handicapped. There are also several hot tubs set at different comfort levels. All the water is purified through Biocarbon filters, ozonation and UV lights. Exercise equipment is connected to generators that produce electricity.

Unlike Grayberg, which imports almost all its food from far away, Greendale produces a great deal of its food right within its city limits, even in winter. There are many small farms in the outskirts, most no larger than an acre or two. None of the farms use pesticides or other toxic substances, and each is a thriving business that provides local employment. Some of the farms specialize in outdoor seasonal crops such as carrots, onions, cabbage and squash, as well as berries and fruits. Others specialize in salad greens and herbs, growing them year-round in greenhouses and extending the production outdoors in spring, summer and fall. The greenhouses are entirely energy-self-sufficient, heated primarily by the sun, with additional heat provided by the chickens, rabbits, pigs, sheep, cows, or horses who live in separate quarters within the insulated greenhouses. The animals are raised in spacious freedom with access to each other, the outdoors and fresh greens, without any of the usual chemicals. They are far happier and healthier than on conventional factory farms. They provide food, fiber and compost fertilizer, as well as carbon dioxide which doubles the greenhouse productivity in winter. Most of the meat that is consumed by the residents and visitors in Greendale is thus produced locally year-round.

Unlike Grayberg, where the school system offers education that to some extent seems irrelevant to living reality, Greendale offers true preparation for good living. The school system was the first to adopt the new solar-dynamic, bio-benign design principles. All schools have solargreen walls that provide heating and cooling, food, bedding plants and tree seedlings. The plants purify the air for the schools and provide wholesome salads for the cafeteria, with excess to sell.

The schools were also the first places to set up solid waste management systems that resulted in 90 percent recycling. These systems have been adopted by the entire city, saving the residents some $100 million a year.

Students of all ages are also participating in the local farming, manufacturing, building and business, and as they get into high school and college many students earn money working after school. Thus many students are well prepared to enter the work force as productive members of the community. Most end up staying in Greendale, because it is clearly the best place to live.

Greendale could have chosen to follow conventional methods, laws, rules and regulations, as Grayberg did. But Greendale realized that such systems violate the laws of Nature and cause stress, waste and pollution, and spiraling costs. Instead, Greendale chose to comply with the laws of Nature by using bio-benign processes for dealing with wastes, to recycle everything, to produce food and fuel locally, and to use the abundant energy provided by the sun (even though this is 50 percent less than is available in an area like Arizona).

As a result, Greendale is now rated as the most desirable city in the U.S. The air, water and food are pure, the environment and surrounding wilderness pristine. There is full employment, and the standard of living as well as the physical and mental health of the residents are the best in the country.

Consequently, crime and social disorder are the lowest anywhere. Money goes around and around in the city, instead of being bled off to far distant places to pay for imported food and energy supplies. The city is exceptionally beautiful, with parks and plantings expanding every year. Arts and culture are rich and varied and available to all.

The wild mountains and the clean beaches, as well as the many innovative solar-dynamic, bio-benign methods, draw visitors from all around the world. Needless to say, people who live here want to stay, and many more want to move here. There is tremendous pressure to expand, and there is some room for expansion up into the foothills of the mountains. This is being done with carefully controlled planning and true public participation, in order to maintain the exhilarating, prosperous, clean and peaceful quality of life that has been attained. The development expertise is available right within the city and is in fact one of the main export items of the city.

Greendale has become a lighthouse that shines bright and clear, a guide to help both large cities and small communities across the country and the world to make livable homes out of their ailing societies.

PART FIVE

A COLLECTION OF POWERFUL QUOTES

These quotes were gathered from many reliable sources. Most of them describe realities that are in actuality far more devastating than any words can describe, more horrible than we can possibly imagine. I offer them as tools to use in whatever way you chose to participate in the good work of righting what we know is wrong.

The element carbon has become one of the largest waste products of modern civilization. During 1988, some 5.66 billion tons were produced by the combustion of fossil fuels - more than a ton for each human being. Another 1 to 2 billion tons were released by felling and burning forests. Each ton of carbon emitted into the air results in 3.7 tons of carbon dioxide (CO_2). Thus, at least 24 billion tons of CO_2 entered atmosphere from these processes in 1988 alone.
United States, is by far the most carbon intensive country in the world. With less than 5 percent of the world's population, the U.S. causes 24 percent of the world's carbon emissions, at more than 5 tons of carbon per person, compared to the United Kingdom at 2.73, Italy at 1.78, France 1.7, Mexico 0.96, China 0.56, Indonesia 0.16.

From Worldwatch Institute Paper 91, Oct 1989: "Slowing Global Warming: a Worldwide Stragedy", by Christopher Flavin.

❦

United States emitted almost 1.5 billion tons of carbon (5.2 billion tons of CO_2) in 1995 from fossil-fuel burning, gas flaring and cement manufacturing, 5.6 tons of carbon (20.7 tons CO_2) per U.S. resident.

Oak Ridge National Laboratories, 1997

❦

The single biggest volcanic eruption in modern times was Mount Tambora in the Java Sea in 1815. It blasted about 100 million tons of carbon dioxide into the atmosphere. In comparison, by burning fossil fuels, human beings are now putting as much carbon dioxide into the air, each year, as one hundred Tamboras. *From "The Next One Hundred Years", by Jonathan Weiner*

Dense clouds of pollution are turning Earth into a gray planet. It is much heavier than during my first space flight ten years ago," says Paul Weitz, commander of the space shuttle Challenger on its maiden flight in March 1983. " Our environment is apparently fast going downhill. We're fouling our own nest. The crud level gets higher and higher. *Chicago Tribune, April 24, 1983*

ꕥ

Lighting uses about a fourth of all electricity used in the U.S., consuming the energy produced by 120 large powerplants (about 4/5 directly and 1/5 in extra airconditioning energy to remove unwanted heat). By using the most efficient sources of electric light in the most efficient ways, and by capturing more of the daylight reaching our homes and businesses, we can profitably reduce our eletricity consumption by up to 90 percent.

Each CFL (compact fluorescent lamp) we replace for an incandescent bulb prevents the emission of 1,000-2,000 pounds of global warming carbon dioxide from powerplants, and 8-16 pounds of sulphur dioxide that causes acid rain. Each CFL also eliminates the need to produce and dispose of up to a dozen incandescent bulbs. In addition, each CFL saves you roughly $25-50 over the lifetime of the bulb. As Amory Lovins puts it: "This isn't just a free lunch, it's a lunch you're being paid to eat!"

The Rocky Mountain Institute building in Snowmass, Colorado uses no fossil fuels, about one-tenth of the amount of electricity used in comparable buildings, and about one-half the water. There is no sacrifice of comfort or convenience. All savings are achieved by the use of solar energy, excellent insulation, and efficient electrical devices and toilets. The net additional cost of the energy-saving features (after subtracting the savings from not needing a furnace) is on the order of $6,000. Compared with normal local building practices and with the cheapest conventional fuels (wood and propane), the building saves more than $7,100 worth of energy per year. This saving repays its own cost in about ten months. *From Rocky Mountain Institute*

ꕥ

During the first half of 1991, tankers spilled 450,000 tons (110 million gallons) around the world, 10 times the amount spilled by Exxon Valdex in Alaska.

Transportation consumes 63% of total oil used in the U.S.

The Arctic National Wildlife Refuge in Alaska is estimated to contain roughly 3.2 billion barrels of oil, approximately the amount the U.S. consumes in 6 months.

The total energy consumed by U.S. agriculture per year is equivalent to more than 30 billion gallons of gasoline (714,285,000 barrels). This represents more than 5x the energy content of the food produced.

The American Farm, by Maisie and Richard Conrad

❧

292 million barrels of oil per year are used to make chemical fertilizers, while millions of tons of manure from cows, poultry and swine are left to pollute the nations groundwater and surface waters.

❧

Like high-input agriculture, genetic engineering is often justified as a humane technology, one that feeds more people with better food. In both cases, nothing could be further from the truth.

Monsanto has patented cotton seed containing genes for Bt (Bacillus thuringiensis). Advertised as being effective against bollworms without the use of additional pesticides, 1,800,000 acres in five southern states were planted with this transgenic seed in 1996, at a cost to the farmers of not only the seed itself but an additional $32-per-acre "technology fee" paid to Monsanto. Heavy bollworm infestation occurred in spite of the special seed, forcing farmers to spray expensive insecticides anyway. Those farmers who wanted to use the seeds from surviving plants to replace the damaged crop found that Monsanto's licensing agreement, like most others in the industry, permitted them only one planting.

Monsanto also manufactures rBGH (recombinant bovine growth hormone) which is injected into lactating cows to make them yield more milk. This is done despite our nation's milk glut and despite the fact that rBGH will probably accelerate the demise of the small dairy farm, since only large farms are able to take on the extra debt for the more expensive feeds, high-tech feed-management systems, and the added veterinary care that go along with its use. The substance causes cows to suffer many problems: bloat, diarrhea, diseases of the feet and knees, feeding disorders, fevers, reduced hemoglobin levels, cystic ovaries, uterine pathology, reduced pregnancy rates, smaller calves, and, most common of all, mastitis. Cows treated with rBGH require more antibiotics, which can transmit to the milk, and which then accelerate the antibiotic resistance among bacteria that cause human disease.

From "A Cruel and Transient Agriculture",
lecture by David Ehrenfeld, professor of biology at Rutger's University,
author of "Beginning Again: People and Nature in the New Millenium."
In Harper's Magazine, October 1997.

Price supports, soil-bank arrangements, direct payments, export controls, research-and-development funds, disaster-assistance payments, marketing agreements, tax write-offs - all have been designed to work chiefly to the benefit of the largest, usually corporate, farmers. The Farmers Home Administration underwrites loans every year overwhelmingly for chemical-based, machine-intensive, monocultural, large-scale farms, thus setting the pattern for local banks and credit institutions, and also for equipment and chemical suppliers. And because federal funds have accounted one way or another for between 20 and 40 percent of all farm income since 1955 - easily the largest single income - what the Federal government does is the single greatest element determining the character of American agriculture. *Kirkpatrick Sale, 1986*

❦

The U.S. National Research Council (NRC) estimates that no information on toxic effects is available for 79 percent of the more than 48,500 chemicals listed in EPA's inventory of toxic substances. Fewer than a fifth have been tested for acute effects, and fewer than a tenth for chronic (for example, cancer-causing), reproductive, or mutagenic effects. Pesticides are purposefully designed to alter or kill living organisms, and they have generally received more extensive testing, but there, too, serious gaps remain.

Between 400,000 and 2 million pesticide poisonings occur worldwide each year, most of them among farmers in developing countries. The 10,000 to 40,000 such poisonings that are thought to result in death each year dwarf the 2000 deaths caused by the toxic gas leak at the pesticide manufacturing plant in Bhopal, India.

While an entrenched agrochemicals industry continues to propound the virtues and necessity of reliance on pesticides, the facts cry out for new solutions to pest problems.

As of 1984, USDA had supervised non-toxic IPM (Integrated Pest Management) on 40 different crops, collectively covering 11 million hectares. Farmers have benefited economically. For instance, a Texas farmer had net returns per hectare averaging $282 higher than other cotton farmers. Apple growers in New York and almond growers in California, using IPM techniques, had per-hectare profits $528 and $769 greater, respectively, than nonusers.

From Worldwatch Institute Paper 79, "Defusing the Toxic Threat: Controlling Pesticides and Industrial Wastes" by Sandra Postel.

❦

Without the modern inputs of chemicals, pesticides, antibiotics, herbicides we simply could not do the job of feeding America. Before we go back to organic agriculture in this country, somebody must decide which 50 million Americans we are going to let starve or go hungry.

Earl Butz, former U.S. Secretary of Agriculture.

❦

CHEMICALS LEFT AS RESIDUES ON FOOD CROPS:

On Apples: Benomyl, Captafol, Captan, Chlordimeform, Daminozide, Dicofol, O-Phenylphenol, Permethrin, Phosmet, Pronamide, Silvex, Simazine, Toxaphene, Zineb.

On Tomatoes: Benomyl, Calcium Arsenate, Captafol, Captan, Chlorimeform, Chlorothalonil, Daminozide, DDVP, Dicofol, Folpet, Heptachlor, Lead Arsenate, Mancozeb, Methomyl, Metiram, O-Phenylphenol, Parathion, Permethrin, Phosmet, Toxaphene, Zineb.

From: "A Study of Carcinogenic Pesticides Allowed in Sixteen Foods and Found in Water: Guess What's Coming to Dinner", by Hind and Spink, U.S. Public Interest Research Group, Washington, D.C. 1989. Published in by League of Women Voters, National Voter, Oct/Nov 1989.

❦

In 1977 the Rodale Research Institute began a study on the effectiveness of alternative farming techniques. Employing a system of crop rotation, animal manures, and soil-conserving techniques, they successfully reduced costs by 10 percent, soil loss by 50 percent, and produced crops that equaled or exceeded comparable conventional systems.

❦

United States generated about 266 million tons of hazardous wastes in 1983, more than one ton for every American. About two-thirds of this waste is disposed of in or on the land through the use of injection wells, and lagoons, pits, ponds and landfills, which inevitably eventually leak into the nations groundwater, causing widespread contamination.

From Congressional Budget Office (CBO)

❦

In Florida, it is estimated that at least three-quarters of all homes are treated with pesticides four to six times a year, and half the lawns are treated with herbicides.

❦

"The chemical industry discharges an estimated 68 million pounds of toxic chemicals directly into U.S. surface waters each year, and 1.6 billion pounds into public sewage systems." *U.S. EPA, 1990*

❦

The military dumped radioactive wastes for 2 decades into the National Marine Sanctuary off San Fransisco, damaging the richest marine habitat on the west coast. But 95% of all radioactivity emitted by nuclear wastes came from the civilian sector - primarily from nuclear power plants. In the 90's it was three times more than in the 80's and 20 times more than in the 70's. Despite this increase, not a single one of the 25 nations producing nuclear power had found a solution to the nuclear waste problem that stood up under close scrutiny.

For decades, in their haste to build nuclear weapons, U.S. weapons manufacturers vented nuclear wastes directly into the air or dumped it on the ground, where it found its way into the groundwater. Radioactive wastes ended up in the Colombia River, contaminating shellfish hundreds of kilometers away in the Pacific Ocean. These facilities accumulated some 379,000 cubic meters (379 million kilograms, 833 million pounds) of liquid high-level nuclear waste from reprocessing, which was emitting 1.1 billion curies of radioactivity at the end of 1989. The wastes are stored in steel tanks at the Hanford Reservation in Washington State and the Savannah River Plant in South Carolina. Tanks have a history of leaking radioactive liquids and accumulating internal buildups of explosive hydrogen gas. Although DOE pledged to clean up these facilities (at a projected cost of $300 billion), it has downplayed the severity of the problems to government regulators and Congressional overseers.

From Worldwatch Paper 106, "Nuclear Wastes: The Problem That Won't GoAway", by Nicholas Lenssen.

❦

In Holland, Dioxin and Furan levels in milk from farms near municipal waste incinerators was found to be 3x higher than normal. The milkfat was separated and sent to hazardous waste incinerators to be destroyed.

❦

No matter how far removed from the centers of industrial activity, people are unable to escape exposure to toxic chemicals. PCBs and DDT, for example, can be detected in the soil and in the bodies of wild animals almost anywhere in the world, as well as in people living in regions of the world still untouched by industry.

The latest research on dioxin and related toxins indicates that these compounds are capable of wreaking silent havoc on the endocrine system, the immune system, the nervous system, and reproductive functions of animals at levels of exposure that are perilously close to those encountered by the average American.

WORLD WATCH, April 1993

Dead Beluga whales in the St.Lawrance estuary were found to be so laden with toxic chemicals that they qualified as hazardous wastes.

U.S. industry generates about 90 billion pounds of hazardous wastes yearly. Less than 10% is disposed of safely.

EPA. Reported in Environmental Action, March 1980

Love Canal, on the Niagara river, became a symbol of the problem of hazardous waste dumps in the United States. The four largest dumps hold enough hazardous wastes to fill 10,000 tanker trucks.

In 1990 Australians experienced a 20% increase in UV radiation, and had the world's highest rate of skin cancer. In Queensland, more than 75% of people who had reached the age of 65 had some form of skin cancer.

An EPA study completed in the early eighties found that more than 70 percent of the 80,000 pits, ponds and lagoons containing hazardous chemicals did not have liners to guard against seepage. The geologic settings of nearly half the sites were such that any seepage that did occur would quickly reach the groundwater. All factors considered, 72,000 impoundments - 90 percent of the total - are thought to pose a threat of groundwater contamination.

The clean-up of some 10,000 toxic waste sites in the U.S. is estimated to cost over $100 billion dollars.

Worldwatch Paper 79: Defusing the Toxic Threat: Controlling Pesticides and industrial Wastes, by Sandra Postel

❦

Approximately 65 percent of North America's uranium deposits lie inside Native American reservation. But these reservations have been home to 80 percent of all uranium mining and 100 percent of the processing, largely because reservations fall outside the jurisdiction of most state and federal environmental laws, and reservation residents have no authority to make their own protective regulations.

With such large reserves of the valuable ore, and given the federal government's historical commitment to nuclear development, many Native American communities should by now have become quite wealthy. But thanks largely to federal land managers, whose goal was to provide preferred companies with the cheapest possible exploitation rights, Native Americans have tended to receive as little as 3.4 percent of the market value for uranium extracted from their lands. Native Americans also have the lowest per capita income of any demographic group in America and the highest per capita rate of malnutrition, disease, and infant mortality.

The Navajo community in particular has suffered from cancer, respiratory ailmants, miscarriages and birth defects caused by radiation. In almost all cases, the people who worked in the mines never received protective clothing or medical checkups or even basic information about the risks of exposure to uranium, and virtually no victims have ever gained any type of compensation. To this day, many Native American communities have to live with illegally high levels of lead, thorium, radium and other toxins that have seeped into their water and soil from tailings ponds and processig plants.

Worldwatch Institute, Paper 127: Eco-Justice: Linking Human Rights and the Environment, by Aaron Sachs.

❦

At the turn of the century, when I entered the Army, the target was one enemy casualty at the end of a sword or rifle or bayonet. Then came the machine gun, designed to kill by the dozen. After that, the heavy artillery, raining death by the hundreds. Then the aerial bomb, to strike by the thousands, followed by the atom explosion to reach the hundreds of thousands. Now, electronics and other processes of the sciences have raised the destructive potential to encompass millions. And with restless brains they work feverishly in dark labs to find the means to destroy all at one blow. *General Douglas MacArthur*

❦

In the United States, about 70 percent of all public R&D (Research and Development) outlays goes to the military. In Israel, about the same, in France about 30 percent, in Italy, Canada and Argentina about 8 percent.

From Worldwatch Paper 89, Michael Renner, 1989

❦

In the 90's, children's TV shows contained an average of 26 acts of violence per hour, up from 18 acts of violence in the 80's.

❦

The U.S. military budget is roughly $350 billion per year, more than one trillion dollars for three years. To put this into a perspective that our minds can grasp I offer the following quote found by William Sloane Coffin in an airline magazine:

"Let's talk a trillion. For one trillion dollars, you could build a $75,000 house, place it on $5,000 worth of land, furnish it with $10,000 worth of furniture, put a $10,000 car in the garage and give all this to each and every family in Kansas, Missouri, Nebraska, Oklahoma, Colorado and Iowa. Having done this, you would still have enough left to build a $10 million hospital and a $10 million library in each of 250 cities and towns throughout the six-state region. After having done all that, you would still have enough money left to build 500 schools at $10 million each for the communities in the region, and after having done all that you would still have enough left from the original trillion to put aside, at 10% annual interest, a sum of money that would pay a salary of $25,000 per year for an army of 10,000 nurses, the same salary for an army of 10,000 teachers, and an annual cash allowance of $5000 for each and every family throughout the six-state region - not just for one year, but forever."

....the principal threat to our future come less from the relationship of nation to nation, more from the relationship between ourselves and the natural systems on which we depend.

Lester Brown, Worldwatch Institute

❦

In the Philippines, 39 children filed a lawsuit calling on the environment minister to cancel all timber licenses. The youngsters said they were acting on behalf of their own generation and those in the future. 37 million acres of virgin forest had already been destroyed, leaving only 2.1 million standing. Devastation of the same magnitude is happening in Russia, in the northwest and northeast USA, in Sweden, as well as in Africa, South America and New Guinea.

❦

Along with the possibility of the extinction of mankind by nuclear war, the central problem of our age has become the contamination of man's total environment with substances of incredible potential harm - substances that accumulate in the tissues of plants and animals and even penetrate the germ cells to shatter or alter the very material of heredity upon which the shape of the future depends.

Silent Spring, Rachel Carson

❦

So important are insects and other land-dwelling arthropods that if all were to disappear, humanity probably could not last more than a few months. Most of the amphibians, reptiles, birds, and mammals would crash to extinction about the same time. Next would go the bulk of the flowering plants and with them the physical structure of most of the forests and other terrestrial habitats of the world.

From Diversity of Life, by E.O. Wilson

❦

If we do not change the direction we are going, we are likely to end up where we are headed. *Chinese Proverb*

❦

Some of the changes needed will be relatively simple to implement. Others will be more difficult. But all will require the courage to see things as they are, to avoid deceiving ourselves, to train ourselves to recognize when sophisticated imbecilities are substituted for serious analysis.

Typically, we cite hugely inflated estimates of the expense involved in changing our current policies, with no analysis whatsoever of the expense associated with the impact of the changes that will occur if we do nothing.

When future generations wonder how we could go along with our daily routines in silent complicity with collective destruction of the earth, will we, like the Unfaithful Servant, claim that we did not notice these things because we were morally asleep?

From Earth in the Balance, by Al Gore

Have we fallen into a mezmerized state that makes us accept as inevitable that which is inferior or detrimental, as though having lost the will or the vision to demand that which is good?

Rachel Carson

A sustainable society is one that satisfies its needs without jeopordizing the prospects of future generations.

From Saving the Planet, Worldwatch Environmental Alert Series

If we want to realize the American Dream we must first wake up.

If it were left to us - architects, builders, the entire construction industry - to set things right, there would be no hope at all.

Malcolm Wells, Gentle Architecture, 1981

Never doubt that a small group of thoughtful, committed citizens can change the world. Indeed it is the only thing that ever has. *Margaret Mead*

❦

The darker the epoch in which we live, the more we must love it, penetrate it with our love, until we have displaced the heavy matter standing in the way of the light which shines from the other side. *Walter Rathenau*

❦

The truth shall set us free.

❦

Go within to find out.

❦

Go placidly among the noise and haste, and remember what peace there may be in silence. As far as possible, without surrender, be on good terms with all persons. Speak your truth quietly and clearly, and listen to others, even the dull and ignorant; they too have their story. Avoid loud and aggressive persons, for they are vexatious to the spirit. If you compare yourself with others you may become vain or bitter, for always there will be greater and lesser persons than yourself.

On the wall of an old church in a small Scottish village.

❦

Never before have the potentials for humanity been so great, nor have the dangers ever been so extreme. *Peter Russell*

❦

Whatever you do may seem insignificant,
but it is very important that you do it.
Gandhi

❦ ❦ ❦

PART SIX

CONSTRUCTION AND MAINTENANCE OF THE SOLVIVA GREENHOUSE AND FARM

What I Did and What I Reccommend

This section of the book offers detailed information and recommendations to help anyone who wishes to achieve high-yield, year-round, humane, organic food production on a home, community or commercial scale, without causing pollution or depletion of resources. I will focus primarily on the construction and management of the Solviva greenhouse and the production of salad greens.

Originally I thought that tomatoes would be the main crop of the Solviva greenhouse. They were spectacular, but they were a lot of work and soon started to develop fickle problems that made the crop difficult to depend on, even after I got the ammonia under control. But I soon realized to my surprise that the real star performer was the new salad product I launched.

"Mesclun", or mixed young salad greens, had apparently been marketed in France since way back, but this was not known to me at that time. I started offering ready-to-eat, multi-variety organic salad that first winter of the big Solviva greenhouse, 1984, for the simple reason that I did not want to decapitate the gorgeous plants in their prime. I was apparently the first in America to market this product, and I called it "Solviva Salad".

The chefs were thrilled. They had never before seen ready-to-serve mixed varieties of salad leaves. When the first chef wanted to know the price of Solviva Salad, I responded by holding up a 1-ounce serving and asking: "What would you charge for a serving like this?" His response was: "At least $4.95". Then I asked: "If you could get this delivered to your door twice a week, clean and ready to serve, what percentage of that price do you think would be fair to pay to Solviva Company?" His reply: "At least 25 percent." My response: "That's about $1.24 per 1-ounce serving. Shall we say $1 per ounce, or $16 per pound. Add to that an edible flower or two, delivery and a dab of light dressing, and your total food cost per salad serving is $1.24. At a menu price of $4.95 you can thus make the normal 300 percent markup."

The price of $16 per pound was at first shocking to the chef, because he was used to paying less than $1 a pound for crates of heads of this and bunches of that. But he knew how much labor and space he had to devote to storing, sorting, washing and draining all those salad ingredients, and how much had to be trimmed off and added to the already burdensome trash removal costs. He realized that this product could make salads the least labor/space-consuming profit center in the kitchen, instead of the most. He immediately ordered 10 pounds, and as soon as he had a chance to experience the convenience in his kitchen and receive rave reviews about the quality from his customers, he upped his orders to 20 pounds twice a week. Other chefs followed suit, all amazed at the beauty and superior flavor, and soon I had requests for more than I could produce.

Soon I heard of "baby greens" being produced in Berkeley, California, in a different way - a vacant lot converted to a gorgeous salad-garden quilt right in the city. Next, a grower in Canada, having heard about Solviva Salad, followed suit, getting $24 per pound, and selling all the way to Boston. Within a few years mesclun mixes were coming in by the ton from California, and the market price fell to $10, then $9, even $5 per pound (no greenhouses needed, and deplorable wages paid to migrant workers). Not long ago, Big Companies started putting out bags of shredded salad mixes at extremely low prices. Along the way the quality of ready-to-serve salad greens has seriously declined, to the point that in many cases better nutritional value can be had from buying the regular heads of lettuce and bunches of greens.

Chefs and other clients continued rating Solviva Salad superior, but I had to lower the price to $14, $13, $12, to $10 per pound to restaurants, and $10.50 to supermarkets. I don't know how far down the selling price will need to go, but in current projections I calculate at $9 per pound. At that price the income is still good enough to make a Solviva farm a good enough business, which of course it has to be in order to survive in the real world.

Nine dollars per pound comes to 56 cents per 1-ounce serving. Add 10 cents for flowers, 10 cents for delivery, and 15 cents for dressing, and the total food cost for a 1-ounce serving comes to 91 cents. These salads are still selling for $4.95 or more in most restaurants, which means that the markup is over 540 percent, well above the average food markup, and way above the markup for such items as steak and lobster. Surely $9 per pound is a more than fair price to pay to the provider of the product. But still, over the years a few chefs have abandoned Solviva Salad because they were able to get something that looks similar from California for a dollar or two less.

For sales through stores, the retail price for Solviva Salad has been $6.95 per half-pound bag. Most people realize that this is a very fair price to pay because the bag contains eight good servings at only 87 cents per serving, and the shelf life can be one to two weeks without a single leaf spoiling. People love the convenience of just pulling out a ready-to-eat handful instead of having to deal with bunches and heads of this and that which must be washed and drained. Yet, some people think it is an expensive luxury food and would rather pay less for an inferior product imported from 3,000 miles away.

There is one more aspect of Solviva Salad I must elaborate on: its health-promoting qualities. So many people have come up to me when they learn that I am the "Solviva Salad lady", and with the deepest sincerity and gratitude testified to the miracle the salad performed for them. In one case a woman's long-standing severe stomach pains went away when she started eating Solviva Salad. People with cancer regained their appetite and energy after chemotherapy. People with AIDS, chronic fatigue, chemical sensitivity, asthma or allergies have told me Solviva Salad made them feel better. Many children who never liked salad before taste this one and then want it every day.

There is a great deal of research that needs to be done to find out just what this food provides. Of all the foods we eat, fresh salad greens have had the most recent and most complete contact relationship with our planet and the sun. I believe that the myriads of living cells contained in each leaf will impart into our body cells extraordinarily life-enhancing chlorophyl, enzymes, hormones, minerals, vitamins, life force and so many other qualities yet to be discovered - but only if the plants are grown with minimal stress and good living compost and good water, and if they are harvested young but not too young, and are ingested before any visible sign of reduced vitality. In addition, I believe that different varieties of greens impart different quantities and varieties of these health-promoting elements, for the same reason that different varieties of greens look and taste different from each other.

If such salad greens were served to patients in hospitals, what effect would it have on their recovery and quality of life? If served in school cafeterias, what would be the effect on the learning and health of the students? If served in prisons, what would be the effect on inmates' attitude and return to society? In all these institutions, salad greens could be grown right on site, in outdoor gardens that would beautify the surroundings and in greenhouses incorporated on roofs and walls. Such greenhouses would also reduce the heating bills and purify the air. Patients, students and inmates respectively could assist the professional gardeners, and that in and of itself would be a healing activity.

I am hoping that the price for Solviva Salad will never have to drop below $8 to $9 per pound. At that price I do believe a Solviva Farm can be a good business, capable of offering good wages to all staff and providing a good return on investment. By contrast, most current farms generate extremely low income per acre and work-hour input, and thousands are going bankrupt.

The level of economic or production success that I believe is possible has not yet been achieved on Solviva Farm. There is the story of what was actually achieved and the story of what I therefore believe is possible to achieve if it is managed with consistent, business-like efficiency.

During the first few years of the business I was responsible for managing the production, and hiring, training and supervising employees and apprentices, as well as marketing and business. (I had no previous experience with any of these factors.) I was completing one construction detail after another, solving various problems that came up (such as the ammonia), and learning from scratch without any role models. I also kept written and photographic records of everything, and fulfilled requests to lecture, consult and write about Solviva. Furthermore, I was continuing my long-standing commitment to designing and promoting comprehensive recycling, solar design and bio-benign wastewater management for this community. Needless to say, I was wearing too many hats, and this was not conducive to achieving the highest possible production. Nonetheless, the production and income exceeded my highest hopes, again and again ever higher.

For the first eight years the Solviva Salad production occupied less than one-fifth acre, constituting the 3,000-square-foot greenhouse and the 5,000-square-foot outdoor garden. Normal expectations of how much money you could hope to gross from any food production on one-fifth acre are miserably low. I had at first hoped to gross $20,000 to $25,000 a year by producing superb-quality organic tomatoes, herbs, greens, vegetables and eggs year-round. After paying the cost of loans, taxes, insurance, seeds, feed for chickens and rabbits, packaging, and assistants, I hoped that

$10,000 to $15,000 would be left over for me. Not much to be sure, but just about everyone thought even that was wildly optimistic.

However, soon the gross income was up to $50,000 a year, astonishing everyone, myself included. Furthermore, it was clear that if there had been consistent and efficient management, the income would have been far higher. For instance, we were harvesting around 80 pounds of Solviva Salad and thousands of nasturtiums and other edible flowers per week, in the darkest, coldest, shortest days of winter, producing an income of over $6000 per month from this 3000-square-foot greenhouse. If it had been run continuously with business-like consistency, the annual income from the greenhouse and the summer garden could have been well over $100,000.

Reporters descended from near and far to see Solviva, and they proceeded to tell the good news in magazines, newspapers, television and radio, all over the world. People who knew something about the topic said that this was the most energy-efficient greenhouse in the world, with the highest productivity per square foot. Media exposures were followed by floods of inquiries and visitors. Although gratifying, this seriously reduced the time I could spend on proper management.

It is imperative for anyone who hopes to make a living this way to understand that the whole operation must remain the absolute No.1, full-time priority. It is also important to know that there should be at least two main people fully involved, one responsible for continuous high production, the other responsible for marketing, sales and business. I know from difficult experience that one person trying to do both those essential jobs is bound to fail.

Although I loved working in the greenhouse and the garden, in my mind I continued to evolve what I had come to recognize as my lifework: to invent, design, develop, teach and promote better ways to live. I imagined solar-dynamic, bio-benign design for homes, schools and other buildings, in rural and urban locations in various climates. The seeds for this book were germinated early on.

In addition, two problems developed that made it impossible for me to continue running the farm. One was the return of an old back injury, reawakened by the lifting and bending necessary in farming. The other problem was new - a severe skin reaction on my hands. I tried dozens of salves, lotions, creams and ointments, and used gloves and other protection, but found nothing that could prevent my skin from cracking open in deep painful fissures. Apparently I had developed allergies to substances in one or more of the plants or possibly in the soil. Now that I have reduced my gardening activities to just home-scale, I can keep both the back and the skin problems to a low hum by taking proper precautions. But let this be a warning to anyone who is contemplating making a living by gardening: back and skin problems could constitute a serious occupational handicap.

Thus, after I had proven my point - that high-yield, year-round production of high-quality food is possible in a climate like New England without any heating fuels, cooling fans or pesticides - I wanted to greatly reduce my involvement. I continued to train assistants in order to free myself as much as possible to pursue the larger vision. But the less time I spent hands-on, the less productive the farm was, and the gross income never did reach much beyond the $50,000. Although $50,000 on one-fifth acre is considered phenomenal, it is obvious that so much more can be achieved.

I finally sought people to take over the whole management of the farm, people with the interest to keep it top priority and the skill and efficiency to bring it up to its production and income potential. A wonderful young person applied. Not much experience, but it would be a dream come true for him. He learned fast and everything looked promising. Then came the summer, and, because the production was way up, marketing had to be done. He got several new accounts, but while he was off marketing, the whole continuous production cycle collapsed due to inadequate watering, reseeding and weeding, because he did not have a production manager or reliable staff. As a result the production and income plummeted. That manager became discouraged and left, and another enthusiastic promising young person appeared, as if heaven-sent, but he too eventually lost the continuous production cycle.

Throughout these painful roller coaster rides, there were brief periods of a month or two in which production would soar to previously unprecedented heights, sometimes 350 pounds per week for a month or more in the summer. Even at such times a high percentage of growing beds was unproductive because they were full of weeds or bolted plants. It was obvious that this farm, with outdoor gardens by then expanded to almost 10,000 square feet, could be far more productive. I now project that a one-acre farm, managed the Solviva way, can produce over 50,000 pounds per year of the highest quality organic salad greens per year, plus herbs and edible flowers and some 50,000 organic eggs and 4,000 pounds of organic chicken and rabbit.

The illustration that depicts a one-acre Solviva farm, 270 by 160 feet (one acre is 43,560 square feet), shows a 10,000-square-foot Solviva greenhouse, which includes processing areas and the indoor animal areas, as well as a small store and cafe. In addition, the acre contains almost 5,000 square feet of low-cost walk-in Solviva Growsheds, plus over 28,000 square feet for outdoor beds, walkways and outdoor animal areas.

With optimal management, the greenhouse and cold frames could produce some 20,000 pounds of Solviva salad for the eight cold months of October through May. That is 320,000 superb one-ounce salad servings, or an average of 9,300 servings per week. If this can fetch a wholesale price of $9 per pound, then the income for those eight months would be $180,000. If the production and processing are done efficiently, 2 pounds can be produced per work-hour. Thus it would require about 10,000 work-hours to produce 20,000 pounds during the eight cold months. For the four summer months, June through September, with the outdoor gardens in addition to the greenhouse and the growsheds, the production can be about 34,000 pounds. That comes to 544,000 one-ounce servings, or 32,000 servings per week. At $9 per pound this results in a gross income of $306,000. At the same production rate of 2 pounds per work-hour, this would require some 17,000 work-hours.

Looking at the whole year together, the total production could thus be 54,000 pounds (864,000 one-ounce servings), for a total gross income, at $9 per pound, of $486,000. Since my favorite scenario also contains a salad bar and retail store, a certain percentage of the salad will bring more than $9 per pound, and since the sales will also include organic eggs, chicken, rabbit and lamb, the total projected gross income for a one-acre Solviva Farm could be well over $500,000 a year.

The actualization of these projections is obviously dependent on highly efficient and consistent management. Might it be difficult to find managers and assistants with sufficient skills and commitment? Yes, it may be difficult. When I was the active, day-to-day captain of the ship, I had some marvelous assistants. Together we would work with joy and efficiency, through heavenly or hellish weather, carrying on deep spiritual communication, singing, laughing and joking, without missing a beat with our hands and eyes. But, for the various reasons that I have explained, I was neither willing nor able to continue being the full-time captain, and to the extent that I was not, production, quality and income were reduced. I have tried to find a replacement, or preferably a team of two. So far I have been unable to find the team to bring Solviva Farm up to its potential. Plenty of temporarily eager seekers, but no team, and no sustained commitment.

I recommend that everyone, including the management, be paid a modest basic salary and that profits be divided among not only the investors and management, but also all the staff. Thus about 50 percent of the earnings would be in the form of profit sharing. There is probably nothing so effective as a good incentive program to keep the production and quality high.

A one-acre Solviva Farm as depicted may cost $300,000 to $350,000 to build to turnkey completion, including the blueprints and consulting needed to enable success, and I believe this investment could bring the good return required from such a high-risk business venture. The actual cost will vary widely depending on climate and labor cost.

A Solviva Farm can be adapted to work effectively in almost any climate, even in urban areas, without animals. Let's get a little visionary here and calculate what it would take to provide five salad servings per week to 7 million people in New York City: 7 million people consuming five servings per week amounts to 35 million servings per week, or 1.82 billion servings per year, which, divided by 16 servings per pound, amounts to 113.75 million pounds per year.

Since I project that one acre can produce over 800,000 superb salad servings per year, it follows that about 2,275 acres would be required to produce 1.82 billion servings. This acreage can be found in small sections on roof tops, south walls and empty lots all over the city. Thus New York City could become salad self-sufficient.

Rooftop and south-wall greenhouses would keep buildings cooler in summer and warmer in winter, thereby reducing the current cost, pollution and depletion that are caused by conventional heating and cooling. Attached greenhouses would purify the air for people living or working in the buildings, and the plants would benefit from people-generated CO_2.

This food production would also provide thousands of good jobs, as well as overall improved health in the city as a result of better nutrition and air quality.

By comparison, rough calculations reveal that current methods of providing the same number of salad servings entail shipping some 225 million pounds (twice the poundage needed because about 50 percent is wasted before it reaches the plates) of salad materials from California, 3,000 miles away. Shipping these greens requires 30,000 refrigerated trailer truck loads, consuming some 20 million gallons of diesel fuel (emitting some 400 million pounds of CO_2). The normal 50 percent waste ratio results in more than 100 million pounds of wastes. These wastes could be transformed into some 30 million pounds of good compost, but instead most of it ends up as garbage at the Freshkills landfill on Staten Island.

THE SOLVIVA DESIGNS

Review, Evaluations and Recommendations

The Foundation

I want to begin this section with one of the few designs that failed, the foundation, and to describe what I recommend doing instead. The foundation has to serve three major tasks: provide support for the building, keep out pest animals such as rats and racoons, and provide insulation. The foundation for the first Solviva greenhouse has successfully supported the building through several hurricanes and torrential rains and blizzards, but it failed to keep out pest animals, and after these animals dug and chewed their way through, the insulation value was greatly reduced.

This foundation consisted of concrete posts set on footings 8 feet apart, 3 feet deep. An insulated foundation beam was built on top of the posts, and then a "skirt" was suspended from this foundation beam down to the footings all around. The skirt consisted of a "sandwich" of two layers of 1-inch mesh chicken wire, 2-inch styrofoam, and two layers of 10-mil reenforced poly fabric.

I thought this skirt would successfully provide both good insulation and an animal barrier. I was wrong. It provided great insulation the first winter, but by the second winter various pest animals had managed to chew their way through, and no matter how often we plugged up the holes, the animals, motivated by wonderful memories of warmth and grain inside, soon worked their way back in. First it was rats, who not only ate the animal grain, but also dug tunnels and burrows in the growing beds, under the water tub, and in the wall and roof insulation. They have caused an enormous amount of damage over the years. Rat bait and various types of traps have only slightly reduced their relentless onslaught. The rats were soon followed by skunks and racoons who were equally destructive.

It is essential to build a ratproof foundation. A solid wraparound foundation wall deep enough to prevent animals from burrowing under might be adequate. However, due to my firsthand experiences with the persistence of these pest animals, I would be unwilling to trust that any foundation wall could be deep enough to prevent the animals from eventually working their way under and in. Therefore, I recommend a complete slab foundation, what's known as a "rat slab". First, remove the full depth of topsoil, plus 2 to 4 inches of subsoil, from the entire greenhouse footprint. Then dig down for footings under the perimeter stem wall, the waterwall, the support posts, and under the water tub. Lay reenforcement wire, tied together to prevent any cracks through which even tiny animals could possibly enter. Then pour a solid 4-inch slab. Such a foundation will raise the cost of construction, but in the long run it will save money, time and peace of mind.

The Wood Framing

I was advised not to use wood for the framing of a greenhouse, that, unless it was cedar, cypress or pressure-treated, it would rot within a few years. Yet, I was drawn to wood, for many reasons - economics, local access, aesthetics, ease of construction, and the possibility of making the adjustments and improvements that would inevitably be required in an innovative design process. Many old wood barns are still in good condition after a hundred years or more, and I felt that if this greenhouse were built well, the frame would not rot.

I chose regular untreated construction lumber for almost the entire frame. Only for the sill beam (which I thought might be frequently wet due to condensation from the glazing) and the main support posts and crossties that actually sit in the soil did I chose pressure-treated wood. Because of the toxic chemicals (copper, chrome, arsenic) impregnated in such wood, I wanted to minimize its use to those areas where the untreated wood would certainly rot. The pressure-treated wood was wrapped in plastic fabric to prevent the toxins from being absorbed by the soil or the roots.

After 14 years, in spite of extremely high humidity in the greenhouse over prolonged periods, none of the framing has rotted. For the purpose of maximizing the light, the south 2-by-6-foot rafters that support the glazing, as well as the collar ties, were painted white. I used ordinary white interior undercoat paint, and it is still as white as it was, with no mildew or mold. This is astonishing to people who know wood, mildew, rot and greenhouses. Of course, 14 years is not very long, and the question still remains of how long the wood will last.

Most standard greenhouses use metal framing, which may well be more practical in the long run, but I have no experience with this.

The Ventilation

Thirteen vents, each 1 by 2 feet, run along the south knee wall, and another 13 vents, each 4 by 2.5 feet, along the top of the north wall. Two doors and one large upper window on the west wall, hinged on the northern vertical edge, scoop in the prevailing southwest winds. Two doors and a large upper window on the east side let out the hot air. Another door on the north wall sucks in cool air. There are no cooling fans.

This ventilation system has proved adequate in this location. The first few summers the greenhouse was filled with basil, melons, cucumbers, and peppers, which all grew with record speed to gigantic size and with exceptionally fine flavor. After Solviva Salad became the primary crop, the greenhouse was filled with edible flowers and many varieties of succulent greens. They have been thriving within the greenhouse even through the worst heat waves. Since the greenhouse in summer is generally a few degrees warmer than outside, all the lettuces, notoriously intolerant of heat, are planted in the outdoor summer garden.

The hot air rushes by its own thermal power out through the top vents and east window and doors, and cooler replacement air rushes in through the bottom vents and the west doors and window and north door. Only under the relatively rare conditions that combine high heat and sun with strong northerly wind does it get close to being too hot. This is because the hot air that wants to exit through the top vents, which face north, is blocked by the oncoming north wind. Under such circumstances a light overall showering of the plants provides significant evaporative cooling.

In summer the vent lids are propped open with sticks. For cold-season venting we installed automatic vent openers, and they worked fine for a while. They function by means of heat swelling a gel that is contained inside a tube. The swelling gel in turn pushes a rod that opens the vent lid as much as 12 inches. No electricity is required. It was exciting to see the vents begin to open precisely when venting was needed, around 9 a.m. on a cold sunny day, close down when clouds passed over, and open up again as the sun reemerged. But soon a serious problem developed: strong northerly winds exerted a lot of pressure on the large north-facing vent lids, which in turn put too much weight on the automatic vent openers. This caused the seal in the tube to break and the gel to leak out. Thus there was no longer sufficient gel to push the rod to open the lid. The danger with automatic devices is that when they malfunction and no one is around, the damage can be devastating.

It is clear from my experience that it is possible to achieve high productivity in a greenhouse, even under the hottest conditions, without the high cost and pollution caused by the huge fans that are required in most modern greenhouses. Under no circumstances should the ventilation openings be reduced, not even in the coldest climate, but I do recommend a few improvements to make it work even better. For instance, the new Solviva designs, for both the greenhouses and the homes, have a ridge ventilation system that opens to both north and south. Thus the hot air will be able to rush out unrestricted no matter which direction the wind is coming from. There are dampers inside the base of this ridge ventilation system, much lighter-weight than the vents on Solviva greenhouse No.1, and protected from the pressure of oncoming winds or snow. These dampers can be controlled with either heat-operated openers or an electric thermostat.

The new Solviva designs also have larger vents on the bottom south wall, as well as more openings on the north wall.

The Paths

The walking paths through the greenhouse are simply composed of subsoil - no concrete, gravel, or wooden planks. The paths work well and are easy to rake clean. A lovely little groundcover plant, baby-tears, has spontaneously sprung up along the edges.

The main path, running east-west, is just over 2 feet wide, which is a bit narrow for a cart or for passing another person, especially carrying a harvest basket. I now recommend making it 3 feet wide.

The side paths that run down between the beds were made just under 2 feet wide, and this we found to be unnecessarily wide. Since you want to maximize the growing area, the paths should be as narrow as possible but still wide enough for comfortable access. They need only be wide enough to fit the legs and feet of one person, so I now recommend 16 inches.

The Growing Beds

We built the growing beds 6 feet wide, but this turned out to require a little too much stretching, especially for someone short. I know from experience how painful it is to pull a back muscle and how long it takes to recover. I now recommend a maximum width of 5 feet, which makes for an easier 30-inch stretch from either side.

The height of the beds is about 14 inches by the main path, which makes it easy to step up on the wooden side ledge of the beds. The path slopes down toward the south glazing so the beds are about 30 inches high at the far south end. The paths were sloped to allow headroom as you get close to the lower end of the glazing. Where the sides of the beds are low, you can comfortably sit on the 6-inch-wide edge board to harvest or plant, and where they are high you can lean up against the sides while you stretch across the bed and thus not strain your back.

To construct the beds, we first set cedar posts into the ground 4 feet apart, keeping the tops of the posts level. We stapled reenforced, waterproof, 10-mil woven poly fabric on the outside of the posts. Then we nailed 1-by-6-inch spruce boards along the sides, level with the top of the posts, with 1 inch of space between the boards, to enable the wood to breathe and dry out between waterings. Finally we laid another 1-by-6 spruce board on top of the posts and the side boards. This forms an excellent T-beam construction, strong enough for two people to stand on at the same time while hanging up the growtubes. The soil on the inside of the beds prevents the beds from collapsing inward, and a galvanized wire running across the beds from post to opposing post prevents the sides from collapsing outward into the paths.

The spruce boards were not treated at all, yet they have not rotted after all these years. I am sure that they would have been long gone if the soil had been up against them, but clearly the poly fabric backing has preserved them.

However, to my surprise, the cedar posts began rotting within a few years, and of course when the posts rot, the sides begin to collapse. We have gradually replaced the cedar posts with locust posts, which will last for a very long time.

I highly recommend both the design and the materials (with locust or other very long-lasting posts) used for the growing beds. They were low-cost, and have proven to be strong, comfortable and productive. Also, unlike some other material that people suggested using to form the sides of the beds, such as bricks, concrete blocks or railroad ties, the boards and posts do not take up any significant amount of space.

The Hanging Growtubes

The hanging growtubes are made with white "schedule 20" PVC plumbing pipes, 10 feet long and 4 inches in diameter, which cost $5 to $6 each. With ear protection and goggles we drilled 1/4-inch drainage holes 4 inches apart along one side of the PVC pipe. Then, with a magic marker, we drew six evenly spaced 18-by-3-inch pockets along the opposite side and cut them out with a jigsaw. We learned from experience that cutting the pockets too wide makes the growtubes too floppy, and that making them too narrow makes it too awkward to fill, plant and empty the growtubes. The ends are closed off with 2-inch plastic packaging tape, so the soil does not fall out.

It takes about five minutes for two efficient people to fill a 10-foot growtube with growing mix, put in 30 young plants, and hang up the growtube on two suspended ropes. The ropes are dacron, nonstretchy and extremely strong, with four strong 18-inch strings attached to the rope 12 inches apart, and an S-hook tied at the end of each string. In a quick easy motion the strings are wrapped around the growtube, and the S-hook looped around the thick rope. The first (lowest) growtube hangs about 2 feet above the growing bed, and, with 12-inch spacing between each of the four growtubes, the fourth (top) one is about 5 feet above the beds. Thus, by standing on the 6-inch-wide top board of the beds, you can reach the top growtube for harvesting and watering, and you can also reach under the bottom growtube for working in the bed.

Strong and lightweight, these growtubes have worked extremely well, capable of sustaining high yields of lettuces for two months or more, before needing to be replaced with new seedlings. Other plants that have proven highly productive in the growtubes include endive frisee, mache and arugula. These growtubes have just about doubled the production capacity of the greenhouse.

I highly recommend these growtubes, though there certainly is room for improvement. They would work even better if they could be emptied by just flipping them over, rather than having to scoop the dirt out. (For high efficiency, every moment and every movement counts.) I dream of a system in which you pull into the greenhouse a 6-foot-long worktable/cart with grow-mix, transplants, and a tub for the spent grow-mix, and 5-foot growtubes (manageable by one person) come to you on hanging conveyors, at the touch of a lever.

I hope to find a better material than PVC for these growtubes because although light-weight, durable, recyclable, and nontoxic in use, the pollutants from the current manufacturing processes are harmful to people and the environment.

The Watering Systems

We installed two different watering systems in the greenhouse, one underground, the other aboveground.

The C.I.T. (Capillary Irrigation Tube) underground irrigation system, invented by Bo Jufors, of Sweden, is truly marvelous, capable of saving time, money and water. The tubes ares buried 12 inches deep, which results in deep, resilient root systems. Unfortunately, this system will not become available in the U.S. until someone decides to become the distributor (which I am sure could be a very good business, considering the severe water limitations in many regions of the U.S. and the many drawbacks of current available irrigation systems).

There is also an aboveground hose system, which works well. Two lines of 1-inch PVC pipe, connected to the water main, run the length of the greenhouse. The top one runs just under the solar water-heating ledge above the second floor catwalk and has hoses 12 feet long attached every 24 feet. These upper hoses can comfortably reach all upstairs areas. The lower pipe is attached to the side of the catwalk and has hoses 20 feet long attached every 24 feet. These lower hoses reach all downstairs areas. This way all areas can be easily watered, without long cumbersome hoses that kink up. The hoses are kept off the walkways by hanging them on hooks.

It is important always to have a good supply of extra hoses and fittings, especially fan sprays, for they begin to leak before long. Some dripping is acceptable so long as water does not drip directly on the plants or cause puddles and mud in the walkways. It is important to turn off the water main whenever no one is in the greenhouse. You can imagine the amount of damage that can be wrought in a short time by a burst hose.

The Solar Heat Storage Systems

The Solviva greenhouse has both passive and active solar-heat storage systems. The passive system consists of 26,000 pounds of water contained in waterwalls both downstairs and upstairs. We built the framework out of ordinary framing lumber. Picture two stud walls constructed with 2-by-4's on 2-foot centers. These two stud walls are set 1 foot apart, connected by 1-by-3's. Two levels of platforms are built into the downstairs waterwalls, totaling 36 platforms, each strong enough to support a bag filled with 400 pounds of water. We attached 4-inch-mesh plastic-coated wire fencing to the outside of this framework, and secured this fencing with battens screwed into the framework uprights. We constructed single-tier waterwalls upstairs. A total of 65 very strong compartments were thus created, placed downstairs and upstairs in what I felt would be the least productive areas, the northeast and northwest corners. Then three large, super-strong plastic bags (not biodegradable!) were evenly inserted, as a triple set one inside the other, into each compartment. We then filled each of the triple bags with 50 gallons (400 pounds) of water and then tied them closed.

This waterwall system has performed better than my highest expectations. After a cold night the temperature of the water in the bags is 55 degrees F (neither the water storage nor the soil ever fell below 55 degrees!). At the end of a sunny day the water temperature has risen to an average of 80 degrees F, more at the upper level, less at the lower level. One BTU ("British Thermal Unit") is a measure that denotes one pound of water rising one degree. Since there are a total of 26,000 pounds of water in the 65 bags and the temperature has risen 25 degrees, that means they have picked up 650,000 BTUs of solar heat. This warmth is slowly released during the night and is about the equivalent to burning 6 gallons of fuel oil.

I highly recommend this waterwall system, but in the new greenhouse designs I have made a great improvement. In the Solviva greenhouse No.1 I placed the waterwalls in the northeast and northwest corners and the animals behind them, because I believed these back corners would be far less productive than the rest of the greenhouse space. But now I know that even these spaces could be highly productive. Therefore, I now recommend that the waterwall run the entire length of the greenhouse, about 8 feet from the insulated north wall, with the animal areas between the waterwall and the north wall. This new design results in a greenhouse that within the same size footprint has more production area, more water storage, and more space for the animals. In addition, it is easier to construct and manage.

The active solar heating and storage system consists primarily of a black pipe coiled inside the peak of the greenhouse. One thousand feet long and 1 inch in diameter, it heats the 600 gallons of water contained in the center water tub. This tub is a 7-foot-diameter, 2-feet-deep galvanized steel farm stock watering tank from Sears, which cost about $230, including shipping. The black pipe starts at the pool and extends up to a wood-framed ledge that we built inside the peak of the 104-foot-long greenhouse. This ledge is covered with black plastic. The pipe, coiled 10 times back and forth, lies on top of this black surface and is covered with clear plastic, and then returns to the tub. A DC pump, powered by the sun through the photovoltaic panels and batteries, was installed to circulate the water through this black pipe. The first pipe that was installed was made of polyethylene, but it started to melt and leak around the joints because of the intense solar heat that was generated. This pipe was soon replaced with a more heat-tolerant polybutylene pipe. This system is capable of heating the water in the tub from its normal winter low of 55 degrees F to 100 degrees after a sunny day. Thus this system absorbs some 200,000 BTUs during a sunny day and then releases it during the night.

There is another active solar water-heating system that contributes a smaller but still significant amount of warmth. This consists of a 12-inch-diameter tunnel that was dug along the north foundation insulation. In this tunnel stand 350 glass gallon wine jugs. (I put a want ad in the paper, and the jugs were contributed by the dozens.) The jugs are filled with water and capped. Air from the top of the greenhouse, about 90 degrees, is distributed down through the tunnel via two poly ducts, forced down by two 16-inch-diameter DC fans. The fans are direct-connected to two photovoltaic panels, meaning that no batteries are employed for this task. When the sun comes out, the fans go on with a terrific force, a powerful and self-evident demonstration of the ability of photovoltaic panels to generate electricity from sunlight. When a cloud covers the sun, the fans stop. When the sun reappears, the fans instantly turn back on. The hot air passes by the bottles, which are spaced about 1 inch apart, and during the course of a sunny day the temperature of the water in the bottles rises from about 55 degrees to about 70 degrees. This system absorbs over 42,000 BTUs on a sunny day, equivalent to about 1/3 gallon of oil. Not much, but it is free and nonpolluting, and every BTU counts on a subzero blizzard night.

The Back-up Heating System

Although I was confident that no heating fuel would be required, I nonetheless wanted to have a backup system - just in case. Thus I created a fire chamber underneath the 600-gallon water tub just mentioned above. Before installing the tub, I dug a trench 7 feet long, 2 feet wide and 18 inches deep. For the purpose of creating a lining for the fire chamber, I cut a 50-gallon oil drum in half lengthwise with a jigsaw (wearing gloves, goggles and ear protectors!). The bottom of the oil drum was left on one of the halves, and a 6-inch hole was cut into it for the chimney pipe. I laid down the two oil drum sections to form a 6-foot-long liner, with the stovepipe cutout at the far end. I packed the ground firmly in an 8-foot circle, and pressed four 8-foot angle irons into the ground, laid across the fire chamber. Then I laid a 7-by-7-foot piece of heavy sheet metal on top, to serve as a hot plate. A 6-inch galvanized stovepipe was installed, leading from the end of the fire chamber and out through the north roof, and the tub was set on top of the hot plate, then filled with water. A roaring fire in the fire chamber, fueled by only paper, cardboard, and kindling, is capable of heating the 600 gallons of water in the tub from 55 degrees to over 100 degrees in less than two hours.

The first winter, when the greenhouse was filled with 100 huge tomato plants producing thousands of superb fruits, turned out to be a record-breaking cold one, with almost no sun for weeks on end. Since the greenhouse was not yet quite finished, this backup heating system was a tremendous blessing. It is highly effective, very simple and low-cost, takes up no space, and plants can thrive right next to it. As with almost all the other Solviva greenhouse designs, this system worked better than my highest expectations. Once the greenhouse was finished, this backup heating system never needed to be fired up again. Occasionally, though, we would fire it up just for the joy of a deep long soak, for this makes the greatest hot tub.

The Earthlung Filter

The purpose of the Earthlung filter is to remove the ammonia from the air in the chicken room so this air can be brought into the plant room, thereby providing beneficial CO_2 to increase the productivity of the plants.

We dug 2 feet of topsoil out of the 16-by-4-foot bed next to the chicken room and lined the sides of the bed with airtight plastic fabric. Then we put in 4 inches of coarse gravel, and on top of this laid down two parallel perforated pipes, capped at the ends. This was topped with more coarse gravel, to a total depth of 12 inches. We covered the gravel with porous nonrotting landscape fabric, and then topped it off with a 12-inch-deep layer of leaf mold mixed with sandy soil. Then we hung four branches of 4-inch flexible ducts within the chicken room, joining them into one duct containing a small DC fan powered by the sun through the photovoltaic panels and batteries. Then we connected this with the ducts laid in the gravel. Finally, the Earthlung filter bed was planted with wheatgrass and Chinese cabbage. These plants grew with tremendous productivity, but because this bed received a lot of ammonia, which led to high nitrate levels in that soil, we realized that food crops may end up with levels of nitrate too high for human consumption. We therefore instead planted the bed with nasturtiums to attract both beneficial and harmful insects.

This filter has now performed its job reliably for years, totally eliminating the problem of ammonia damaging the plants, while greatly raising the CO_2 level of the air in the plant room and thereby increasing plant productivity. The only maintenance needed is to once a year skim off most of the 12-inch layer of leaf mold, which has by then turned into nitrogen-rich superior compost, and then replace it with a new mix of leaf mold and poor sandy soil and replant it.

In the latest Solviva greenhouse designs, the Earthlung filter is much larger, installed in the whole length of the bed in front of the waterwall, which in turn has been enlarged to run the whole east-west length of the greenhouse.

The Attached Cold Frame

We constructed a cold frame, 3 feet wide, along the whole length of the greenhouse, up against the knee wall below the south glazing. This was for the dual purpose of adding another 300 square feet of production area and to reduce the heat loss from the greenhouse. It had two sets of lids, one double-glazed with Sungain glazing to let in the light and keep in the warmth, the other, made of plywood, strong enough to handle the snow load that would come crashing down the south glazing, as well as strong enough to make it possible to shovel off the snow.

Yet another job for the lids, when they were open in the summer, was to act as a wind scoop to increase the ventilation coming in through the bottom vents. This job they performed very well.

But we had to remove the cold frames because of two serious drawbacks. One problem was that, although crops grew quite well in there, it was a backbreaker to work in because we had to kneel on the ground and reach way in to plant or harvest. But a much more serious problem was caused by the fact that, because the roof of the cold frame was built at a shallow pitch, the snow did not slide off. Therefore, the snow built up to over 5 feet high on top of the cold frame and the lower half of the greenhouse glazing. This proved to be too much stress on the glazing system.

The Glazing System

For a greenhouse to be highly productive through cold, dark winters it needs a glazing system that admits as much light as possible, over 90 percent. The glazing also needs to provide excellent insulation. The four layers of 3M Sungain polyester glazing on the Solviva greenhouse No.1 perform these dual requirements admirably, insulating as well as four layers of glass, yet admitting as much light as only one layer of glass. This glazing material is also extremely strong, as has been proven by the hurricanes, hailstorms and snow loads it has had to endure. In addition, it is very long-lasting, rated to last 10 to 15 years without significant reduction in strength or light transmittance.

The south roof consists of twenty-six 2-by-6-inch rafters, 24 feet long, spaced 4 feet apart and painted white. Posts, set 8 feet apart, support a beam that runs east to west, which provides support for the rafters at mid-point.

The first layer of glazing was applied, with 3M Isotac double-stick tape, on the outside of the rafters. Layer No. 2 was applied on the inside of the rafters, adhered with the same tape and also by 1-by-1-inch wood spacers screwed into the rafters. Layer No. 3 was taped to the inner edge of those spacers, and then a second set of 1-by-1-inch spacers was attached, with layer No. 4 applied to them with only the double-stick tape. Thus this glazing system contains three airspaces, one about 4 inches plus two about 1 inch each. This application was not easy. Working with large areas of floppy glazing and sticky tape, in the wind and with mostly inexperienced volunteers, it was not possible for us to get the job done with the sort of thoroughness that was required to attach it tightly enough to prevent dust and small insects from getting in between the layers.

A vacuum pump was installed at each end of the greenhouse to create negative air pressure between layers No. 1 and No. 2, to prevent the outer layer from flapping in the wind. This tremendously reduces stress to the plastic film and consequently slows down the aging process. The pumps suck out the air in a serpentine pattern through 2-inch holes drilled alternately at the top and the base of each rafter, 6 inches from the ends. This ventilation may explain why there has been no sign of mildew or rot on the rafters.

The inner three layers of glazing are now 14 years old, and still in excellent condition, except that, because the installation was not tight enough, some dust and bugs did get in between the layers. The outer layer did not fare as well. The Isotac tape gave way under the tremendous pressure of 5 feet of snow which accumulated on the lower half of the glazing. This happened because snow sometimes fell heavily in the middle of the night and piled up on the cold frame covers, and thus on the lower half of the glazing, before we had a chance to shovel it off. When the Isotac tape no longer held the edges down, the glazing started flapping wildly in the wind. Each time, as soon as conditions permitted, we refastened the glazing, but it sustained serious deterioration. Needless to say, the cold frames were soon removed.

In spite of the brutal stress of record-breaking snowfalls and many incidents of violent storms, snow, ice, and hail - including the infamous Hurricane Bob which leveled thousands of trees and utility poles on the island - the glazing lasted for a long time. I have no doubt that if it had been applied correctly to start with, it would still be fine.

Sungain glazing was developed to be placed between two layers of glass, as a tri-pane or quad-pane window. But even though this turned out to be cost-effective in the long run because of the tremendous energy savings, it was outflanked in the market by low-E coated glass that appeared about the same time. Thus, even though Sungain proved to be a superior glazing, 3M stopped manufacturing it a couple of years after the Solviva greenhouse was built, before we had a chance to convince 3M that there may be a future for greenhouses. Therefore, this wonderful glazing material is no longer available. Over the years I have explored what other superior glazings may be available. It seems that the solar heating research and applications that surged under President Carter's leadership in the 70s deflated like a punctured tire under the Reagan administration in the 80s. Thus, the prevailing opinion nowadays is that glazing should minimize the intake of solar energy in order to reduce the cooling load of buildings, which is being accomplished by varying degrees of tinting the glazing. I believe it is much better to maximize the solar gain and to prevent overheating by means of outventing the heat or storing it, as need dictates.

Two companies, DuPont and Hoechst, still make glazings that have the same high light transmittance as Sungain, plus life expectancy in excess of 15 years. I cannot testify to either because I have not yet had a chance to try them, but they both look promising, though more costly than Sungain. There are of course the standard greenhouse glazings, but they have problems that vary from low light transmissivity to short life spans.

The new Solviva greenhouse designs have a new glazing system consisting of separate panels that are assembled on a work table, with two to five layers of glazing, depending on the climate. I believe it will be easier to make the glazing dust- and bugproof this way instead of working overhead with large floppy sheets of plastic. These glazing panels are then screwed into a lighter-weight greenhouse framing.

The Pitch of the South Roof

There is widespread controversy about what is the best roof pitch for a greenhouse. I thought long and hard about what pitch would be best for the south roof, the most important factor being that it must be capable of shedding snow. If it was built at a shallower pitch, say 35 degrees, snow may remain on the glazing for weeks at a time. Although snow makes excellent insulation, it also would prevent the light from reaching the plants, and this would cause great harm within a day or two. If on the other hand the angle was unnecessarily steep, the interior growing space would be reduced. I settled for a 45-degree pitch, and this has proven perfect.

The Walk-in Grow Shed

This is a very simple 200-square-foot light-weight wood-framed structure, 25 feet long by 8 feet wide. The peak is only 7 feet high, and a 16-inch-deep and 18-inch-wide trench down the middle provides full headroom under the collar ties. A 3-foot-wide bed runs on either side of the walkway. A door at each end, plus top vents, provide good ventilation to keep it cool enough even through the hottest days without any fans. It is a great place to raise seedlings April through October. Cold-tolerant salad plants, such as kale, radicchio and mache, remain productive in there throughout the winter without any heating. The north, west and east walls can be insulated, and strategically placed plastic gallon water jugs and the ground provide thermal mass to store solar heat. For the proposed Solviva one-acre farm I recommend 19 such growsheds, 32 by 8 feet, for a total of almost 5,000 square feet.

Cost of Materials

The total cost of construction materials (excluding the glazing) for the 3,000-square-foot Solviva greenhouse No. 1, plus the 1,000-square-foot uninsulated barn along the north side, and the 200-square-foot growshed, was around $20,000. This comes to $5 per square foot if you divide it by the total footprint of 4,000 square feet. The ratproof slab foundation that I consider essential would raise the cost. This does not include the cost of any labor, which in this case was supplied primarily by a flock of angels. Obviously I cannot estimate what labor might come to for another project, because it would vary widely depending on location and circumstances. A rough estimate of the full cost for the 10,000-square-foot greenhouse that is designed for the projected one-acre Solviva farm would be about $25 per square foot, or $225,000, including labor. The total cost of the one-acre farm, with a 10,000-square-foot Solviva greenhouse, plus 5,000 square feet of Solviva walk-in cold frames (19 @ 32 by 8), fencing, all equipment, tools, supplies, animals, soil amendments, blueprints and consulting would be somewhere around $300,000, not including the cost of the land.

Building A Greenhouse in a Very Cold, Dark Place, Such as Alaska

If I were building a greenhouse in extreme cold, cloudy and dark conditions, I would add more insulation and more animals, and install a total of five layers of highly light-transmissive glazing. In addition, I would add deep solar-heat storage mass underneath the greenhouse floor, installing the "rat slab" 4 to 6 feet down, and bottle up solar heat during the summer months. With sufficient CO_2, salad greens would do quite well with significantly shorter days than in New England. In the extreme north, with no sunlight in mid-winter, I would sow seeds under lights in a small separated section of the greenhouses, transplant the seedlings into six-packs in a slightly larger area, and time it all in such a way that they didn't need to be planted into the permanent greenhouse beds and growtubes until sufficient light returned.

❦ ❦ ❦

SOLVIVA GREENHOUSE MANAGEMENT

It is not possible to convey in a book all the knowledge that is required to be successful. For those who are seriously interested in producing food the Solviva way, there are blueprints and consulting available. It has always been my intent not only to invent and develop the Solviva solar-dynamic, bio-benign designs and methods, but also to pass on the know-how to others. During the years that I managed the farm myself, I offered internships, apprenticeships and workshops to people from around the world. Such live-in, hands-on experience is clearly the best way to learn, and I hope Solviva Farm will someday again be able offer this. I do so hope that some good people will recognize the great Solviva business potential for right livelihood (doing well while doing good) and swoop in to restore the greenhouse and farm and reactivate the mission to produce the best products and best learning experience.

I want to start this section by describing my experience with insect management, because the primary reason for the prevalent use of toxic chemicals for food production is pest insects. No matter how well you manage the greenhouse, insects will be there to a greater or lesser extent.

Insect Management

Insects are some of the most fascinating creatures on Earth. Some are so sensitive that they can smell food or a member of the opposite sex miles away. They can relay complex messages to each other and can perceive and react instantly to environmental threats. According to E.O.Wilson (the famed Harvard entomologist and author of the book Diversity of Life), of the 1.4 million species of life forms thus far identified on our planet, 751,000 are insects. Wilson adds that recent research and exploration, especially in the world's tropical rainforests, has revealed that there may be 10 million, or many more, additional species of insects still left to be identified.

Ninety-nine percent of all insects are considered beneficial to mankind. They eat harmful insects, pollinate flowers, and aerate the soil. They make compost out of dead plants and animals, which in turn breaks down the minerals in the soil, making them available as nutrients. Insects make silk, honey, wax and shellac. They provide food for birds and fish and each other.

Insects with a rapid reproductive rate, such as fruit flies, are also used extensively for advancing our understanding of genetics, heredity and the effects of pollution on ourselves and our environment.

However, about 1 percent of all insects are harmful to us, and we wish we could exterminate them to the point of extinction. These are the insects that eat our homes and our clothes, our trees, bushes and flowers, our crops and our foods in storage, and the insects that spread disease, sting, bite or suck our blood.

Since ancient times people have tried various methods of protection against insect pests, but in the last 50 years many have become worse in spite of, and indeed because of, the prevalent toxic techniques of human intervention. When modern pesticides were first developed in the 1940s, they seemed to be a miracle cure against pest insects. However, it soon became apparent that some of the insects survive the sprays because they are resistant to the poison, and that most of their offspring inherit that resistance. Thus the insect infestation soon bounces back, this time more virulent because most of the beneficial pest-controlling insects are destroyed along with the pest insects, and they are slower to recover. Ever more toxic pesticides are invented to try to keep up with the rapid evolution of the pest insects. This results in great damage to the whole ecosystem. I believe we cannot win the battle against harmful insects through toxic methods.

In early 1960 Rachel Carson wrote Silent Spring with dire warnings about pesticides. Yet, the EPA estimates that since then use of pesticides has increased twelvefold, while the percentage of crop losses before harvest has doubled. More than 2 billion pounds of some 320 different toxic pesticides are manufactured annually in the United States (just counting the "active" ingredients), of which over 1 billion pounds are applied in the U.S., while the rest are exported.

Some of the most toxic pesticides are banned in the U.S. but are still manufactured here and then exported to countries with inadequate or nonexistent regulations and controls. There they are applied by unprotected workers to grow cash crops for export to United States. Strawberries, bananas, tea, coffee and sugar are produced by giant corporations all over the world on land previously used for growing food by and for the native populations. Barrels used for shipping the toxic chemicals end up being used to store drinking water and food. Pesticide Action Network reported that worldwide there are some 25 million pesticide poisonings annually, 220,000 resulting in death.

I am still appalled by the U.S. policy reflected in the following news clip from about 1981: "A United Nations resolution outlawing the export of dangerous products without the knowledge and consent of importing countries, passed in the General Assembly 146:1. The lone "NO" vote was cast by U.S. representative Jean Kirkpatrick, who said the resolution would harm American industry since it is the largest exporter of hazardous materials".

I do not believe that her "NO" vote reflects the will of the American people.

Most of the American food supply, whether imported or grown in the U.S., contains traces of hundreds of pesticides. So does the water we drink and the air we breathe. It is said that there is no place on planet Earth, no land, no water, no air, no matter how remote, that does not contain detectable levels of pesticides. In most cases the levels of individual toxins are below what has been determined to be hazardous. However, research has shown that some of these pesticides become many times more toxic when combined with others in the food chain, in the soil, the air, and in the groundwater. Moreover, some have proven to be more carcinogenic and mutagenic when tested on lab animals that are fed low-protein diets, alcohol or antidepressants, and recent findings show strong evidence that pesticides suppress the immune system, reduce human sperm count, and tamper with the endocrine system.

This evidence is very real and very frightening, yet some people believe that if we do not continue to develop and apply herbicides, insecticides and fungicides, food will be scarce and unappealing and there will be unprecedented famine and economic devastation in the U.S. as well as the rest of the world. I believe that the opposite is true.

Here I hope to help dissolve some of the mystery and fear surrounding insects in general and pests in particular, while providing information on how to control the pest insects, not with the conventional methods involving poisons, chemicals, gas masks, complicated tools and machinery, warnings, dangerous side effects and high costs, but with simple, low-cost, safe, effective, harmonious management techniques.

The insect management techniques that I practice in my home and in the Solviva Winter Garden greenhouse have evolved over almost 40 years of gardening, and have gradually become more and more effective, simple and reliable. When I first began to grow food indoors in my home in the late 70s, I wanted to control insects without any toxic pesticides for the simple reason that I did not want to use anything that might be poisonous to people. I read everything I could get hold of on the subject and talked with people who had experience with organic gardening, eagerly trying many of the techniques suggested.

I used rotenone until I found out that it was highly toxic not only to insects, but also to me as I was applying the powder or spray. It does lose most of its toxicity within three days of application, but it is actually a dangerous nerve toxin when fresh. The last time I used it was years ago, dusting it on sheep to kill sheep keds, and it gave me a violent headache.

I tried many methods recommended in books and articles. For instance, I collected aphids by the thousands and ground them up in my blender to make "bug juice". This took a lot of time, was not very pleasant, and did not get rid of the aphids. I used insecticidal soap and oil of cedar in various spray contraptions that kept clogging up. I collected horsetail, a primitive plant prevalent since the time of the dinosaurs. My blender got a real workout as I followed directions for liquifying the horsetail plants to extract their rich silica content. I then sprayed this liquid on the plants. The silica was supposed to form crystals to increase the vigor of the plants so they would more effectively be able to resist aphids. The silica did form crystals which covered the plants, but this did not reduce the aphid population significantly, and I did not like the way the white powdery crystals covered up the beautiful green leaves. These methods were minimally effective for controlling the pests, and I had a vague feeling that even these mild substances were in some way upsetting the balance of nature.

Besides aphids there were many other insects, most of which I could not identify. I knew that some, like ladybugs and praying mantis, were "good bugs", and others, like aphids and scale, were "bad bugs". But I also saw eggs, larvae, pupae and adults of little creatures I could not identify as either "good" or "bad". One time I saw a little cluster of 10 to 15 very fine hairs about 1/2-inch long protruding straight up from the top surface of a hibiscus leaf. Each hair had a tiny light green pinhead-sized egg or spore or whatever at the end of it. Was it friend or foe? I searched through my books but had no luck finding anything like it. I finally decided that it looked most like a virulent fungus. So in order to prevent it from spreading to the whole plant, I nipped off the leaf and discarded it. Later I found out that I had destroyed the eggs of one of the most valuable beneficial insects, the green lacewing.

Another time I found a whole bunch of little larvae crawling around on a nasturtium plant close to the ground. One-quarter-inch-long, spool-shaped, black with orange dots on the sides. Again, friend or foe? I found that none of the books described or pictured all the various developmental stages of any of the insects. If these larvae were vegetarians, they could cause devastation as they multiplied. If, on the other hand, they were carnivorous, they just might control the aphids.

I got myself a 10-power magnifying lens and got down next to the little creatures. I was going to find out through my own observation. I picked a leaf that was covered with aphids and nuzzled it up under one of the little larvae. It soon stepped onto the leaf, and a moment later I observed through my lens what I have since seen many times: this unknown creature was actually picking up an aphid in its jaws and eating it!

Now I knew it was a friend, and I proceeded to carefully distribute these larvae to various plants infested with aphid colonies. Within a short period of time the aphid population had significantly diminished. I soon realized what these little unknowns were. I saw a larva attached to a post transforming itself into a pupa, and then the pupa slowly bursting open to reveal a full-fledged ladybug, not black with orange dots like the larva, but the normal reverse. Every day for over a week I observed a cluster of 1/8-inch bright orange-yellow, football-shaped eggs, standing up in tight formation on a leaf. Finally one morning I had the honor of observing miniature ladybug larvae hatch out, one through the top of each egg, and crawl down the stem in search of food. One of the funniest sights was a ladybug rapidly scooting around with her mate clinging to her back for what seemed like a very long time. She busily explored and munched on aphids, while he, I assume, fertilized the eggs within her.

I pored through all the books on insect management that I could borrow or buy. I took workshops on integrated pest management (IPM), and followed the advise of planting "biological islands" with flowers and herbs, importing beneficial insects from producers, and spot-spraying with the least toxic stuff like insecticidal soap.

As I observed the various harmful and beneficial insects in close interaction with each other, I increasingly felt that they themselves had the balance all figured out. Spraying even with the "nontoxic" stuff felt like a rude and mistrustful intervention. My spraying techniques, which were already limited to just the areas with significant pest populations, became still more narrow in focus, until I found myself just aiming at individual aphids in an attempt to avoid harming beneficial organisms. Finally even that felt like an insult to the beneficial insects who had come to do their job. So I stopped spraying altogether.

I have compared notes with those who use various minimal or least toxic spray IPM techniques, and it seems to me that my zero-spray methods are not only simpler, but also more effective. I now consider the zero-spray approach to be one of the most important keys to successful insect management.

From from time to time there are rushes of aphids or whiteflies or spider mites. I hang on faithfully to my trust in harmony, allowing some of the plants to be overcome by these undesirable vegetarians, watching in suspense and cheering on as the beneficial insects of amazing variety increase in number.

At the end of the first winter in the Solviva greenhouse, 1984, a teacher of horticulture from the local high school came to visit. The high school greenhouse was being managed with the standard chemical fertilizers and pesticides. From time to time there would be a sign on the greenhouse door that read "DANGER! KEEP OUT!", and the place would reek of some nauseating poison. This is how this teacher had learned to control pests, and this is what he was teaching the students. He was highly skeptical of my experimental methods, and when he came to visit there was an impressive, almost depressing aphid rush in progress. "Now what are you going to do?" he asked rather smugly.

During the next couple of weeks I watched spellbound as the aphid population increased, followed closely by the rapid increase in numbers and varieties of beneficial insects. I had ordered some of these from California, and others that I had never seen before had apparently developed as part of the existing ecosystem within the greenhouse.

Then one morning, as I did what had become my daily insect inspection routine, I couldn't believe my eyes. There was hardly an insect in sight! What I observed was a most dramatic example of the beneficial insects suddenly getting the upper hand over the harmful insects, consuming them all. And once the feast was done, the beneficials had left for better hunting grounds. Since it was by now early May, they left the greenhouse. I had never seen so many beneficial insects in my outside garden as that summer, and my greenhouse was inoculating the entire neighborhood for the benefit of all.

The teacher came back for a visit soon after "The Big Disappearance". He slowly wandered throughout the greenhouse, closely inspecting, mumbling to himself in awe: "I can't believe it...with no pesticides!?...I can't believe it..."

I do so wish I had a video depicting that sequence of events. Nothing quite so dramatic ever happened again, nor did I ever again have such an aphid boom. I subsequently learned to introduce appropriate beneficial insects long before the situation became so critical.

While weeding, watering and harvesting, it is easy to check for insects and redistribute beneficial ones to places in most need. As we wash the perfect-sized individually picked leaves from all the 50 or so varieties of salad plants, we return any beneficials we find to the greenhouse. Occasionally an insect will slip past our scrutinizing eyes and end up in the bags and boxes of Solviva Salad. I encourage people to see them not as something disgusting, but instead as living proof that no toxic substance has ever touched these greens.

Viewed through an 8- or 10-power magnifying lens, an extraordinary world is revealed. The endless varieties of mandibles and proboscises, waving antennae, ovipositors, long jointed legs and compound eyes, gossamer wings and pearlescent and metallic colors are more fantastic and beautiful than one can ever imagine without having experienced it. What natural processes can create the pure gold that adorns the chrysalis of the monarch butterfly? And why?

I marvel at the delicate sea-green beauty of the lacewing as it flits around with rapidly moving wings, like a little fairy. Or the syrphid fly, hovering like a hummingbird as it seeks nectar in tiny flowers. Its plump little larva sits on a leaf gobbling up an aphid like a sea lion feasting on a fish. A tiny orange Aphideletes aphidimyza caterpillar siphons the body fluids from an aphid, reducing it to a shrivelled white husk.

I watch in utter fascination as dozens of tiny braconid caterpillars emerge out of the sides of the dying body of a much larger leaf-eating caterpillar. The little braconids perform belly dance gyrations as they each immediately spin themselves a cocoon. The cocoons remain piggybacked in a cluster on the host caterpillar for a few days as it slowly crawls along, no longer a threat to the plants. A couple of days later as I watch the little bundle of cocoons, a tiny hatch suddenly opens at the end of one of them and out comes an adult braconid wasp, 1/8 inch long. After a few moments of waving its antennae and preening its wings, it flies off to begin a new cycle of feeding on nectar and mating and finding an egg or larva or adult vegetarian insect in which to lay its own eggs.

Research is rapidly progressing all over the world to find effective, safe bio-logical, bio-benign controls for every kind of pest, to make it possible to achieve high production levels with all crops without using any toxic chemicals, even for such insect-prone crops as corn, soybeans and cotton. There are still challenges out there, some not related to food production, such as ticks, fleas, wool moths and mosquitoes. For these pests there are some recent developments, such as pheromones (sex attractants) that may be so species-specific that they do not adversely effect the ecosystem.

I believe that to get off the conventional poison treadmill successfully it is best to "go cold turkey", to drop any sprays or dusts that are meant to kill or harm insects (except perhaps a few species-specific ones like BT) and instead put all the effort into a comprehensive program of harmonious management techniques.

In a greenhouse the most important part of that program is to lay a good foundation, to create a thriving soil rich in compost, nutrients, organic matter and a profusion of living organisms. The other parts of the 13 Golden Guidelines complete this foundation.

Whether you are growing food or ornamentals in a home sunroom garden, or a commercial or community greenhouse for feeding hundreds, you have to deal with insects. I hope that the illustrations in this book will open up a window to reveal to you some of the life cycles and interrelationships of most of the various insects that you are likely to encounter in an indoor garden. Remember, though, that as you find insects not identified in this book, nor in any other, and you wonder whether it is friend or foe, it may well be a friend. Observe it with respect and treat it as beneficial until proven harmful.

ꕥ ꕥ ꕥ

13 GOLDEN GUIDELINES

FOR

MINIMIZING PEST PROBLEMS

1. KEEP THE SOIL ENRICHED WITH COMPOST AND A GREAT DIVERSITY OF LIVING ORGANISMS.
2. WATER THE PLANTS JUST RIGHT.
3. MAINTAIN GOOD LEVELS OF HUMIDITY AND AIR CIRCULATION.
4. MAINTAIN HIGH LEVELS OF CARBON DIOXIDE.
5. MAINTAIN TEMPERATURES AS CLOSE TO IDEAL AS POSSIBLE.
6. PROVIDE THE BEST POSSIBLE LIGHT.
7. GROW A MULTITUDE OF SMALL NECTAR-PRODUCING FLOWERS.
8. GROW NASTURTIUMS AND OTHER FAVORED HOST PLANTS FOR THE VEGETARIAN INSECTS.
9. INTRODUCE AND ENCOURAGE A GREAT DIVERSITY OF BENEFICIAL INSECTS.
10. LEAVE SOME INSECT-INFESTED PLANTS STANDING.
11. DO SOME TRAPPING AND HANDPICKING.
12. MONITOR IN ORDER TO KNOW WHAT IS GOING ON.
13. TRUST THAT HARMONY WILL PREVAIL.

(Yes, 13 is my lucky number.)

1. Keep the Soil Enriched with Compost and A Great Diversity of Living Organisms

Keep the soil alive and thriving by adding plenty of good compost made with animal manure and plant wastes. I want to stress the importance of including animal manure, for I believe that compost that contains only plant wastes is anemic by comparison.

Compost is a miraculous substance that constitutes the very foundation of life. Try to conjure up in your mind the following fact: one teaspoon of living compost contains some 5 million bacteria, 20 million fungi, 1 million protozoa, and 200,000 algae. The life and death processes of these miraculous life forms slowly dissolve the minerals in the soil so that they can be absorbed by the minute roots of the plants. Good compost is also essential in order to maintain an ecological balance that will prevent the harmful critters from dominating the ecosystem. Add about 4 gallons of living compost to each square yard of growing area three to four times a year, depending on the crop.

Whenever the surface of the soil begins to look closed or compacted or grows a film of moss or algae, lightly cultivate to a depth of no more than an inch (an ordinary table fork works well among the closely spaced salad plants). This does not damage the delicate plant roots, and it greatly improves the soil's ability to breathe, to inhale oxygen and exhale CO_2, which promotes the health of this miraculous ecosystem. Do this when the soil surface is fairly dry, because handling soil when it is wet is injurous and must be avoided at all cost. It causes soil particles to glue together, constricting the tiny passages that allow oxygen, nutrients and multitudes of microorganisms to flow through the soil.

2. Water the Plants Just Right

Plants do best with thorough watering when they need it and no watering when they don't need it. If you water when the soil is damp, it becomes waterlogged, preventing oxygen from penetrating freely through the pores of the soil. This endangers the earthworms and the whole soil ecosystem, including the roots of the plants, as well as those beneficial insects whose life cycles partially take place underground, such as the gall midge (Aphideletes aphidimyza). Any reduction in soil vitality results in reduction of plant vitality, and therefore makes them less productive and more vulnerable to insects.

In the winter it is preferable to use sun-warmed water (everyone, even plants and insects, prefer a nice warm shower instead of ice-cold). Water can be delivered through an underground irrigation system or through aboveground hoses with soft spray heads that don't harm the leaves or disturb the insect populations.

Wilting should be avoided as much as possible, although it sometimes occurs on a sunny day even when there is enough moisture in the soil. The cure for this is another kind of watering which is an exception to the above rule ("never water when the soil does not need it"). This special watering consists of a very light quick sprinkling of the leaves which does not penetrate the soil. Cold water is actually best for this purpose. This is also good to do routinely on sunny mornings to freshen the whole environment with negative ions and to provide plenty of water droplets for the beneficial insects. It is important not to do this in the afternoon, because mildew might result if the leaves remain wet during the night.

3. Maintain Good Levels of Humidity and Air Circulation

In a small home greenhouse spraying and misting lightly may be required in order to increase the humidity for the benefit of the whole ecosystem (including the people). In a larger greenhouse filled with an enormous area of leaf surfaces constantly exhaling moisture, it is important to keep humidity to a minimum in order to prevent fungus infestation and also in order not to inhibit the plants' transpiration, or breathing, which could reduce their productivity.

One of the best ways to limit humidity is to do all watering before noon on a sunny day, so that the leaves can dry off before nightfall. Another way is to use an underground irrigation system, for this keeps the surface of the soil much drier. Open some vents whenever outside conditions allow in order to let excess moisture escape. Circulate the greenhouse air with small fans, strategically placed to prevent pockets of stagnant air.

4. Maintain High Levels of Carbon Dioxide

Carbon dioxide is an essential building block for plants, without which they cannot grow no matter how perfect the other conditions such as light, nutrients, moisture, or temperature. In a standard greenhouse the plants have depleted most of the available supply of CO_2 building blocks by about 11 a.m., and when there is not enough CO_2, plant growth stops. Transpiration may continue and thus the absorption of water and nutrients through the roots, but the result is plants puffed up with water and undigested nitrate. As stated earlier, it is not good for people to consume food too high in nitrate, because this reduces the oxygen exchange capacity of our blood.

Conversely, if CO_2 levels in the greenhouse are increased above the normal outdoor level of 350 ppm, to a level of some 1,400 ppm, thereby increasing the number and density of carbon building blocks in the air available for the leaves to absorb, the result is increased plant productivity, increased dry matter ratio and decreased nitrate levels in the crops. Increased CO_2 compensates tremendously for the low light levels of northern mid-winters, as more and better growth can take place within each hour of daylight. This results in higher productivity and healthier plants better able to resist insects.

In the Solviva Winter Garden greenhouse the important task of increasing CO_2 levels is performed primarily by the resident chickens and rabbits, by the litter in their dens, and also by the compost in the growing beds.

5. Maintain Temperatures as Close to Ideal as Possible

For optimum productivity the greenhouse temperatures should be kept above a minimum of about 40 degrees F (higher for plants like tomatoes) and below a maximum of about 90 degrees F by opening and closing vents as appropriate. In the beginning I thought that on a really cold, sunny day I would have to keep in every degree of sunheat in order to sufficiently warm up the water mass for the cold night. I soon realized that even if it was zero degrees F outside, on a sunny day the greenhouse temperature would climb to over 110 degrees F if the vents were left closed. This was obviously stressful for plants and insects - and for people. I found out that enough heat was absorbed in the water mass even if I cracked open the vents to keep the temperature below 90 degrees F.

As in so many aspects of the greenhouse management I find it easiest to understand what to do if I empathize with the animals, the plants and the insects: if it feels too hot for me, then it most likely is also too hot for them.

The resident animals (in this case chickens and rabbits, but it could be any animals that appreciate living in a warm space, such as pigs, horses, cows, or goats) give off surprising amounts of warmth (approximately 8 BTUs per hour per pound of body weight), essential under the coldest conditions.

In summertime both the Solviva greenhouse and solargreen home stay comfortable even in the worst heat, without any fans. This is due to the roof ridge vent design that lets out hot air, as well as sufficient vents, windows and doors that create good cross-ventilation.

6. Create Optimum Light Conditions

Salad greens, tomatoes and other food plants are normally grown in the summer and in full sun. Since the winter sun comes in much weaker at a lower angle and for fewer hours per day, with more cloudy days, it is easy to understand that in winter the plants need to receive the very best light conditions possible. The combined layers of glazing, whether plastic or glass, should have a light transmittance of 90 percent or more.

In addition, many plants, especially food plants, dislike seeing light from only one direction. The result of unidirectional light is leaning, leggy, weak plants, less productive and more vulnerable to insects. It is important that the plants receive light not only from the south and above, but also from the north. You might think that this means that all-around glazing is the best, as in standard greenhouse designs, but this results in unacceptably high heat loss. Solviva greenhouse design provides good light conditions in spite of the whole north wall being insulated and opaque. This is accomplished by tilting the north wall at an angle to best reflect the sunlight on to the plants, 60 degrees at this latitude, and by covering the interior surface of the north, west and east walls with white UV resistant plastic. You might think that silver foil or mylar would be more effective. I tried it on one section of the north wall and found that just a small percentage of the surface was brilliantly light, but a far greater percentage was dark gray. This resulted in leaning and elongated plant stems.

With this design I never need to use artificial light (nor extra warmth), not even for raising seedlings in the darkest, cloudiest, shortest days of December. In summer the seedlings are happiest raised in the well-ventilated Solviva cold frames that have one layer of woven plastic glazing which lets through about 80 percent of the light. This protects the seedlings not only from the harsh UV light but also from drenching downpours and whipping winds.

7. Grow A Multitude of Small Nectar-Producing Flowers

Many of the adult beneficial insects live primarily on nectar and do not eat the pest insects. Instead they control the pests by laying their eggs within the eggs, larvae, pupae or adults of the pest insects.

There must therefore be multitudes of nectar-producing flowers in the greenhouse, small enough for these minute beneficial insects to draw nourishment from, such as dill, fennel, parsley, cilantro, marjoram, thyme, alyssum, and different varieties of sage. These flowers are also edible, delicious and attractive in salads, and therefore a valuable crop.

8. Grow Naturtiums And Other Favorite Host Plants for the Vegetarian Insects

This enormously productive plant, with its beautiful, delicious and nutritious blossoms, needs a section of its own because it performs such an important role in the whole ecosystem. It thrives in the less well lighted areas of the greenhouse where few other food plants could be productive. Of all 150 plus different varieties of plants that I have grown in the greenhouse so far, nasturtiums are the most attractive to the greenhouse pests I have encountered in the Solviva greenhouse: aphids, whiteflies, thrips, spider mites and leaf miners. (I have never understood why some people recommend planting nasturtiums as a repellant, because in my experience I have always found it to be a powerful attractant.) And wherever the harmful insects congregate, the beneficials will be sure to follow. Thus nasturtiums provide the primary habitat for most of the insects, leaving the production crops relatively free of insects.

9. Introduce and Encourage a Great Diversity of Beneficial Insects

Some of them arrive on their own from outside, such as the syrphid fly, chalcid wasp, tachinid fly and braconid wasp. Some I bring in from outside, such as Aphideletes aphidimyza. From my favorite producer of beneficial insects, Rincon-Vitova in Oak View, California, I order Encarsia formosa to control whiteflies, predatory mites for spider mites, cryptolaemus for mealybugs, and ladybugs and lacewings for aphids and other vegetarian insects. I recommend frequent inspection of the insect population, with a 8- to 10-power magnifying lens. At first sighting of groupings of harmful insects, order the correct beneficial insects, unless you already have them aboard. Avoid crisis management!

10. Leave Some Insect-infested Plants Standing

This is the concept most difficult for others to understand. When people see an insect-infested plant, they want to pull it out and discard it. But when I point out the eggs, pupae and larvae of the beneficial insects and parasitized harmful insects, they begin to realize that if you pull plants as soon as they become pest-infested, you would prevent the reproductive cycles of the beneficial insects.

Furthermore, as insect-infested plants become weaker, they become even more attractive to the harmful insects than to the healthy plants, thereby tending to leave the healthy plants unharmed. I often leave insect-infested plants in place until the plant is pretty much dead, at which point the insects will have dispersed. Sometimes I will remove an infested plant and carefully remove most of the beneficial insects in their various stages of development and return them to the greenhouse.

11. Do Some Trapping and Handpicking

Slugs and caterpillars can be a real nuisance and are really the only creatures I spend any time controlling. Handpicking combined with beer baiting is a winning combination.

Some eyebrows rise when I go into the liquor store and ask for a case of the cheapest beer, but that's nothing compared to the grimace that follows when I say it's for slug control. I serve the beer in little plastic cups, spaced every 3 to 4 feet, pressed down so the top is level with the soil. You don't see many slugs during the day, but after dark has settled they come slithering out from dark nooks and crannies. They are most attracted to the beer, and many end up drowning in it (probably a pretty good way to die), but many slither back out of the cups. You can patrol with a flashlight and a can of water with a little alcohol in it, picking slugs who are feasting on the beer or the valuable crops of gourmet greens. At times I have collected 150 or more in 30 minutes, and I repeat this trapping and picking for four or five days, after which their numbers and damage will be greatly reduced. This cycle needs to be done about two to three times a year.

Pill bugs, or sow bugs, are mostly scavengers of decaying plant matter. In my experience they do not initiate any damage the way the slugs do. However, pill bugs do not hesitate to consider as fair game any plant nicked even slightly by a slug or a caterpillar, or errant scissors during harvesting, and they can therefore be quite destructive. I was happy to find that they also love beer.

The beer cups need to be emptied within two days, in order to avoid the nasty job of dealing with rotten beer, slugs and pill bugs. Beer-pickled slugs can be served to the chickens or added to the compost.

Ant colonies must be eliminated because ants love to milk the honeydew from aphids and go to great lengths to increase "their" herds of aphids and protect them against predators and parasites. This has been a problem only a few times, and placing a couple of "ant cups" in their pathways took care of the problem quickly. Although this ant control contained some toxic materials, it was contained in a small local area. I therefore do not feel that it interfered with the harmonious management of the whole greenhouse ecosystem. I have since learned that there are various non-toxic ant controls available.

Various kinds of moths lay eggs that hatch into voracious vegetarian caterpillars, especially in summer. I have sprayed Bacillus thuringiensis (BT), a microscopic organism found in soils throughout the world. It is said to affect only the leaf-eating caterpillars, but I cannot help wondering if it also affects the larvae of the braconid wasp that eats the harmful caterpillars. Recently a new way to deal with moths has been developed. A sticky trap is baited with the particular pheromone that attracts the particular pest moth. I do not have much experience with this method, but it sounds promising.

12. Monitor in Order to Know What is Going On

You need to be aware of what is going on in the greenhouse ecosystem, and the best way to do that is to scan the plants while watering or harvesting. I carry my trusty 10-power magnifying lens and check here and there the approximate ratio of "good guys" to "bad guys". I have learned through experience to introduce the appropriate beneficial control insects before the harmful ones proliferate too much. Again, avoid crisis management.

13. Trust That Harmony Will Prevail

You need to trust that harmony will prevail. This means not spraying with anything that might in the slightest way harm the beneficial insects, not even insecticidal soap. If it is meant to kill or harm the pests, it most certainly harms the beneficials also, either directly or as they interact with the poisoned pests.

I have at times been tempted to use "something" against aphids when they begin to get obnoxious, usually in late spring. But upon closely examining the situation, I have seen that there was hardly a colony of aphids that was not also being occupied by one or several varieties of aphid-controlling insects. Any spraying, even the sometimes recommended hard spraying with cold water, would be detrimental to the healthy development of the beneficial insects, and even to the plants.

The bottom line is this: if I help my insect friends and trust them to do their job rather than harm them by interfering with harsh and toxic methods, they will maintain balance in the ecosystem. It is very simple and economical, and it works very well most of the time. Occasionally there will be a minor loss of crops, but to my mind, if you spray with toxic chemicals, your crops will be a total loss.

More Detailed Description and Recommendations of the Solviva Management Techniques

✿ ✿ ✿

Composting

Good compost can be considered the start and the heart of the whole operation. It is miraculous that the waste material from animals and plants, properly combined and processed, is capable of producing such exquisite food for us. Good compost contains many earthworms, pill bugs, millipedes, centipedes, earwigs, mites and other visible life forms, as well as billions of microorganisms. The life and death processes of all these organisms slowly dissolve the minerals in the soil into accessible molecules of nutrients which, together with the molecules of nutrients in the composted plant and animal wastes, are absorbed by the plants' tiny feeder roots.

Many people consider composting difficult, and it is true that many a compost pile ends up just yielding a slimy, smelly, fly-infested mess. Much has been written about composting, about proper proportions, layering, turning, inclusion of this or that special expensive compost starter organisms and other amendments. Many different kinds of composting containers are available through stores and catalogues, some at hundreds of dollars, many of which do not make an attractive addition to the garden, and some of which are not very practical.

My own composting techniques are much more casual and have always resulted in excellent compost, "The Best" according to those who know compost.

My favorite bins are made from wooden pallets, which are usually available free for the taking. Choose those that have closely spaced boards to prevent the compost from spilling out through the sides. Four pallets make a bin 4 feet square, and only three more are needed for each additional adjacent bin. I tie them together, straight and secure, with the strings that I save (I am the penultimate string-saver) from the bales of hay that nourish the sheep through the winter.

Into these bins I randomly empty baskets of weeds, spent salad plants and their grow-mix-filled root balls, as well as the manure-filled bedding from the sheep, rabbits and chickens. Proper proportioning seems to happen automatically because of the whole farm production system. No turning is needed because this compost becomes Earthworm Heaven, and the earthworms are the primary workers who perform the necessary blending and aerating. It is amazing how fast they, in collaboration with the other visible and microscopic creatures, break the waste products down to finished compost, even in the middle of winter. Even though the bins may be covered with snow or frozen on the surface, when you dig down just a few inches, it is steaming hot.

In the winter I recommend that all composting be done inside the greenhouse, in the animal areas or right in the plant room, because of the major contribution of heat and CO_2 that is generated by the composting process.

For home-scale composting, I recommend a very simple, trouble-free, effective system. Get a regular trash barrel. Make a couple of small vent holes close to the top of the barrel, and tape window screening over the holes. Then put in about 12 inches of loose, airy leaf mold, wood shavings/chips and soil, followed by about 2 gallons of manure from cows, horses, pigs, chickens, goats, or rabbits (not cats or dogs), or 2 pounds of ground beef. Cover this with an inch or two of soil. Then put in 1,000 or more earthworms. The manure or meat provides the initial food for the earthworms. Then add your household food wastes, including meat, bones and fat, chop it in slightly, and sprinkle in a little soil or leafmold over each load. There is no need for turning or aerating, as this job will be done by the earthworms. Add a quart of water once in a while for the purpose of keeping the mix moist, as earthworms are happiest with about 70 percent moisture, but make sure not to add so much water that it puddles on the bottom. Secure the lid and tilt the barrel a bit so the compost can breathe all the way down to the bottom of the barrel. Each time you add food wastes, rotate the barrel a quarter turn so a new section of compost can breathe. In the summer, place the barrel in the shade; in the winter, place it in the sun.

In my experience, this method results in very rapid, trouble-free composting of all food wastes all year-round, without any odors or flies. If you notice that composting and volume reduction is not happening effectively in the beginning, be patient. It can take a few weeks for the ecosystem within the first barrel to establish itself for most effective composting. When the first barrel is almost full, top it off with some more manure and soil. Start a second barrel, using some of the compost from the first barrel as a starter mix. For an average household the first barrel will probably be all composted by the time the second barrel is full, and the resulting black gold will make your garden very happy. You will have plenty of earthworms for future barrels, as well as for presents for gardening and fishing friends.

There is more information about composting in the sections on solid wastes and animal management.

Soil Fertitility

In order to maximize the fertility of the soil in the greenhouse, I suggest the following: at the start of construction of the greenhouse, after digging out the soil from the greenhouse footprint, spread it out evenly about 8 inches deep, close by, but no closer than 20 feet. Thus traffic on the topsoil can be avoided while the greenhouse is being constructed. This topsoil can now be developed into superb soil for the beds and the growtubes. If you have very shallow or no topsoil, import the best topsoil you can find.

Send a soil sample to a reputable lab for testing, and then add amendments to adjust for any shortcomings. For an average soil I recommend adding, per square yard, approximately the following: 4 gallons of compost, 4 cubic feet of peatmoss, 1 to 2 pounds of soft rock phosphate, 1 to 2 pounds of greensand, plus agricultural lime in accordance with the soil test.

Spread it out evenly and blend thoroughly by rototilling twice, rake it smooth and level, and then water well enough to soak through to the full depth. The next day sow buckwheat seed, spaced 1/4 to 1/2 inch apart, cover lightly and water it well. Let the buckwheat grow for about two to three weeks, to about 10 to 12 inches high, then turn it under and sow another round of buckwheat. Repeat these buckwheat cycles until the growing beds in the greenhouse are ready to be filled. Thus, the very finest quality topsoil will be in the making while the greenhouse construction progresses.

When the growing beds have been completed, but before the glazing goes on, this superb soil can be piled in with a backhoe, taking great care not to damage the rafters.

In order to maintain highest soil fertility, the soil then needs to be enriched with compost and minerals four to five times per year. This is best done as soon as the final harvest has been gleaned from any given area and the spent plants and their root balls have been removed.

For each square yard of growing bed I recommend adding the same amounts of enrichments mentioned above, except the peat moss. With a trowel chop this into the top 4 to 6 inches of topsoil. There is no need to go deeper because watering and the earthworms and other creatures will do a good job of distributing these amendments deeper down into the soil. This is followed by a good but not heavy watering. After the water has had a chance to soak in to the soil evenly, preferably overnight, you can seed or plant.

For perennial plants that grow in the same spot for years, such as sorrel or tomatoes and many of the herbs, this same enrichment needs to be provided three to four times a year. Apply evenly several inches away from the stem of the plants, so you don't nick the stem (which could start disease), and fork it gently into the top inch of the soil, so as not to injure the roots.

I recommend that every area of the greenhouse and gardens receive a buckwheat treatment once a year, except of course the perennial areas. To do this, first enrich and water the beds as above, then broadcast the buckwheat seeds about 1/4 inch apart. Then cover the seeds with 1/4 to 1/2 inch of soil, pat it down, and then water thoroughly but gently enough so that the seeds do not get displaced or exposed. In warm temperatures the buckwheat will sprout in a couple of days and will grow to 8 to 10 inches in a week or 10 days, slower in cooler weather. (In cold outdoor conditions buckwheat does not sprout at all, but instead rots.) At that stage chop it down into the soil, making sure that every stem is indeed chopped. Stems that are not chopped off will come back to haunt you as weeds. Since the heart-shaped leaves are delicious, nutritious and beautiful, you can harvest them and add them to the salad mix, and just chop the roots and stems into the soil.

After allowing this soft greenmanure to mellow in the soil for just a day or two, the bed is ready to plant. You might think that planting in freshly decomposing plant matter would be harmful to the new roots, but I have never experienced any problems with this method.

For the growtubes, I thought that perhaps I could pull out the spent plants and most of their roots, enrich the remaining grow-mix the same as the beds and put in the next generation of 3-inch seedlings. But I found that the longest period of high productivity was obtained by emptying and replacing the grow-mix between each crop. I evolved ever-easier methods of doing this process of moving, emptying, filling and replanting the growtubes. In my notebooks I have the growtubes hanging a foot apart on revolving vertical loops reaching from the bottom of the greenhouse to the top: you stand in one place with a work table/cart in front of you, and the growtubes come to you at the push of a lever. As in just about every aspect of human designs and techniques, there is plenty of room for further improvements.

To make a batch of grow-mix for the growtubes, blend 6 cubic feet of soil-less growing medium such as Metromix or Promix (basically peat moss, perlite and vermiculite; warning: the mixes vary widely in quality) with 3 cubic feet of compost, plus about one cup of greensand, one cup of rock phosphate and one cup of lime. (This amount is always dependent on your pH value, but here on the Vineyard it is needed because we have acid soil, acid rain and acid groundwater.)

By the time the plants have been living and producing in the growtubes for two to four months, their root system will have filled the whole space and used up most of the nutrients. It is thus best to empty the growtubes after the crop is spent, rather than using the same mix for a second crop. The root-filled mass goes into the compost and thus gets reused a couple of months later.

Further fertility is supplied through foliar feeding, done every week or two with liquid seaweed and/or fish emulsion, or compost tea.

Outside Garden Preparation

The need for rototilling or plowing can be eliminated entirely by proper preparation and maintenance. For starting a new garden, or restoring a poorly managed one, I recommend the following: in early April spread compost and mineral amendments at the same rate as recommended for the greenhouse. Then, as early as the ground can be safely worked (here about middle of April), rototill the entire garden thoroughly. It is of utmost importance that the soil not be worked when it is too wet, because this causes the soil particles to stick together, which greatly reduces the soil's ability to breathe.

Then rake out clumps of weeds and roots, especially grass roots, which will otherwise return with a vengeance.

Next, lay out the beds by marking with strings set 5 feet apart. This represents the center of each bed. I have found 5 feet on center to be the best for avoiding back problems. The paths are 18 inches wide, which is wide enough to enable you to take various positions of crouching, kneeling and squatting (important for minimizing back problems) during planting, weeding and harvesting, while the beds are 42 inches wide, which enables an easy 21-inch reach from each side of the bed. For the greenhouse I recommend 16-inch paths and 60-inch beds. This is because the beds are high enough to allow doing the work standing up rather than squatting or kneeling. When you can stand up and lean against the side of the bed, you can easily reach 30 inches to the center of the bed, and you need only 16 inches for your feet and legs.

Mark the sides of the beds with string and pound in 18-inch-long rot-resistant stakes (locust, plastic or metal), 2 feet apart and 14 inches into the ground. Set 1-by-6-inch untreated spruce fence boards, 16 feet long, along the bed side of the stakes, pressing them into the soil about 2 inches. Then dig out 8 inches of soil from the paths and put it into the beds. The soil in the beds will thus be about 11 inches higher than the bottom of the paths, which makes the beds comfortable to work in. The boards do not have to be fastened because they are held in place by the stakes on one side and the soil on the other side.

I highly recommend these spruce fence boards (untreated) because they are low-cost (about 23 cents per running foot), attractive, and easy to use. Mine have lasted for many years, and when they finally rot too much to do their job of holding in the soil, they become compost. For easy access and watering, I recommend that the beds be made 32 feet long, which is two 16-foot boards on each side. Never walk on the beds, for this compacts the soil.

I advise against the use of raiload ties or pressure-treated wood for the edging of the beds, for several reasons:

1. The chrome, copper, arsenic, creosote and other poisons that are injected into the wood for the purpose of retarding rotting will surely be absorbed by the roots of the crop plants, even if only in minute quantities.

2. Some of the toxins will be absorbed by your skin as you lean against the wood for planting, weeding and harvesting.

3. The ecosystem will be effected in ways that will certainly not be beneficial.

4. Chunky railroad ties rob you of important growing space.

Mulching

For crops that stay in the ground for the whole summer or longer, such as tomatoes, squash, potatoes, corn and strawberries, there is no question that mulching is advantageous. For the small, closely spaced salad plants it is not such a clear choice.

Here are some of the advantages:

1. The soil stays more evenly moist, and the need for watering is reduced.

2. The roots stay cooler, which in the heat of summer increases productivity.

3. Weed growth is greatly reduced.

Here are some of the disadvantages:

1. Unless you are careful in your choice of mulch, you can introduce millions of weed seeds into your garden, which end up sprouting right in the mulch.

2. Slugs and other harmful creatures can be a major problem because the mulch makes a great place for them to live.

3. It is a lot of work to spread the mulch carefully in among the closely spaced small salad plants.

4. As you harvest the salad leaves, it is difficult to avoid also getting the pieces of mulch.

Nonetheless, this is what I recommend: cover the whole garden, paths and beds, with about 2 inches of shredded leaves. You can get a professional landscaper with truck, blower and shredder equipment to suck up leaves from piles at the dump, right through his shredder and into his truck, and then dump them inside your garden fence. This mulch will keep the beds moist and weed-free until you are ready to plant and will also add much valuable organic matter to the garden.

When you are ready to plant or sow, simply rake the leaves off the beds and into the paths. Later, when the plants are large enough, this leaf mulch can be picked up from the paths and tucked around the plants to help reduce weeding and watering, and to keep the roots cool.

Seeding

It is extremely important to protect the seeds from any humidity - a securely closed Ziploc bag works well - until ready to sow, because as soon as seeds have been exposed to even the slightest bit of moisture, they "wake up". From then on they cannot be stored but must be planted or else they will die. Even the palm of your "dry" hand contains a considerable amount of moisture, and for that reason do not put more seeds in your hand than you intend to sow in that session. After the seeds have been sown, whether directly in the ground or in seedling flats, they must be kept moist or else they will die.

Young seedlings are vulnerable to damping-off, a fungus disease that attacks the plants at the surface of the soil and causes the stem to rot. This is most likely to be a problem in greenhouses, perhaps because the glazing material reduces the sun's UV radiation, and UV is a powerful fungicide. So for raising seedlings under glazing I recommend taking the following sanitizing precautions. Soak the seed trays in a barrel filled with water plus 2 cups of bleach. Soak for at least a couple of hours with the lid on. Then lift out the trays and let them drain right over the barrel. This treatment effectively kills microorganisms that can cause disease. Then remove the trays from the barrel and give them a quick rinse. Do not empty out this barrel, but use the same bleach solution again and again. When it no longer smells of bleach, top it off with yet another cup or two of bleach. Always keep the barrel covered, to prevent the bleach from evaporating.

Fill the trays evenly to the top with Metromix 360 or Promix or similar soil-less mix straight out of the bag. (Again, watch out - some of those mixes are no good!) Make sure these bags are kept in a dry place to prevent the mixes from becoming damp and moldy.

For seeding by hand into channeled trays, make shallow grooves in the seeding medium with a ruler, and with a simple little hand seeder tap the seeds in about 1/4 inch apart. Cover lightly, to twice the thickness of the seeds, and pat down so that the seeds are in full contact with the seeding medium. Place the trays in shallow water for a few minutes, until enough water has been absorbed to show dampness on the surface.

Or you can use a seeding machine. I finally got one. It was a big investment for this size operation, and it requires some skill to operate successfully. I used "288" plug trays, with 288 plugs each 3/4 inch square and 1 1/4 inch deep, 12 rows by 24, but I found them too small. I would rather grow the seedlings more robust, so I recommend instead using the "128" plug trays, which have 128 plugs, each 1 1/4 inch square by 2 3/8 inches deep, 8 rows by 16. Cover the seeds lightly and water the same as above.

The machine works very well for almost any size round seeds, but not at all for oblong, pointy seeds such as lettuces. Fortunately, many of the oblong seeds that you will want in your salad blend are also sold "pelleted". Each individual seed is actually coated with clay to form a small pellet that works well with the machine.

To achieve the highest, most reliable production it is important to plant most seeds singly, not letting them drop in by twos or threes. It is far less time-consuming to singulate the seeds at the time of sowing than to thin them at the time of transplanting. As usual, a stitch in time saves nine.

Write down the variety, seed company and date on clean plastic labels, one per tray. This is very important in order to keep track of which varieties are the best and which are not worth growing again. You will also find out which seeds have a reduced sprouting rate and should therefore be discarded.

Ideal sprouting temperatures for most seeds are above 70 degrees F and below 90, but I have also been consistently successful with sprouting in the greenhouse at much lower temperatures without any backup heating. Keep the seed trays damp but not too wet.

Before sprouting, the seedtrays can be kept in dim or no light. But at the very first sign of sprouting, the trays must be put in the very best light available or else the seedlings will become tall, leggy and weak. If you have never experienced it, you cannot imagine how fast this legginess can happen. Most seeds will sprout in two to five days, whereas some varieties take much longer. For instance, parsley takes about three weeks.

Seeds can also be sown directly where they are to grow as mature plants. This is especially suitable in the summer outdoor garden when the cycles are faster and space less at a premium. Prepare the beds same as before, and make grooves 2 inches deep and 3 inches wide, with 4 inches between rows. Whether you sow as separate varieties or in compatible mixes, it is very important not to sow the seeds too densely. If you do, the production will be poor because the plants will grow too crowded and will be stunted and succumb to mildew. There are various wheel sowers available for even distribution of seeds. They work more or less well, depending on the variety of sower and the seed shape. As with other methods, cover the seeds with a light medium about twice as deep as the diameter of the seed. Give a thorough but gentle watering, and never let the soil go dry. On a sunny, windy day, mortal drying can take place within a very short period of time after watering. A strip of plastic, paper or fabric laid over the rows can help prevent rapid drying.

Transplanting

For those seeds that were sown by hand in channeled trays, transplant when seedlings have developed their first two pairs of tiny leaves. The first pair are the nurse leaves, the second pair are the first true leaves. By this time they will have grown sufficient roots so that you are able to pull out a whole row as a loosely cohesive ribbon, yet not such long roots that they are too densely intertwined. In summer this stage will be reached in about two weeks after sowing, in winter, a little longer. Shake the ribbon apart with a lightly quivering motion, and separate each little seedling by gently holding it by one of the nurse leaves. Never hold by the stem, as this can easily cause injury to the main transport channels (you would not hold a little baby by the neck either). Try to retain most of the roots, because the more roots you lose, the more stressed the plant will be and the shorter its life span. Be sure to minimize sun and wind hitting the seedlings during this transplanting process, as wilting will cause stress, and (yes, you guessed it) shorter production time.

Next to you keep a stack of bleach-soaked 21-by-11-inch trays with six-pack inserts, rinsed, and filled with a slightly damp mix of one-third compost and two-thirds Promix or Metromix. Press holes almost to the bottom of each cell, and insert one seedling into each. Tuck it in gently, with the base of the nurse leaves at the soil line. Do not plant too deep, burying the growing- tip of the plant - or too shallow, leaving the stem exposed. (Ah, the Zen of gardening...) If the stem has grown too long to fit straight into the hole, corkscrew it down gently.

After filling the whole tray, set it to soak for 10 minutes in shallow water with 1 tablespoon of liquid seaweed per gallon of water. Set the trays away from sun and wind for about 24 hours so that the seedlings can recover from the shock of being transplanted. Then set them in the best available light, and never let them get too dry or too wet.

Water seedlings with liquid seaweed at the first sign of sprouting and then about once a week.

Planting

Planting seedlings into the greenhouse beds and growtubes can be done any time, rain or shine, because the glazing protects against full UV radiation. But for planting outside the seedlings must be hardened off, because the full force of UV radiation can burn the leaves of plants not accustomed to it, causing serious, sometimes fatal, damage. The easiest way to harden off is to plant on a cloudy day. But since that is not always possible, you can plant in the late afternoon and shield the new plantings with a row cover for the next day.

The seeds that were sown into "128" plugtrays with the seeding machine will grow to be seedlings about 3 inches tall in about three to four weeks in summer, and four to five weeks in winter. Let them stay in the trays until the root ball is well filled and holds its shape when removed, but not so long that it gets root-bound and thus stressed. The plugs can be popped out of the trays with a special machine made for this purpose. The machine is expensive, but it is really essential because of the time and plastic trays it saves.

The seedlings growing in six-packs will be 3 to 4 inches tall two to three weeks after being transplanted from the hand-sown seedling trays in summer, three to four weeks in winter. Again, let the root ball fill out enough not to fall apart when you take it out, but not to the point of being root-bound. Remove the transplants gently by squeezing the sides of each cell and very gently pour/pull out the young plant, making sure that the entire root ball comes out.

Once the young transplants are out of their plastic trays, keep them covered with something like a damp old white sheet while they await planting. Space the plants about 4 inches apart. Thus each square foot of bed can hold nine plants. Each 10-foot growtube can hold 30 plants, and, because the growtubes offer so little space for the roots, make sure they are filled all the way to within 1/8 inch of the top. Obviously, spacing depends on what size leaves you want to harvest and also on variety. Romaine can be spaced closer because it grows tall and narrow, whereas tah tsai grows wide and therefore needs wider spacing. There will be space around each plant after each harvest, but the leaves will have grown to the point of overlapping a bit, but not crowding each other, before the next harvest. If they are too closely spaced you may have mildew problems, and also it becomes more difficult to harvest without injuring the leaves.

I made hole spacers that are very convenient. They are constructed from 1/4-inch plywood, 10 by 20 inches, just a bit smaller than the plug trays, with dowels cut and shaped just a bit larger than the plugs. The dowels were screwed and glued onto the board, spaced in accordance with the needs of the different plants. Then I gave them several coats of white "BIN" paint so they would not rot and so they would not retain soil. These boards are pressed quickly all the way into the prepared soil and then removed. In just a minute you can thus make perfectly spaced and shaped holes for a whole bed, then follow with a plug tray, pick out four to five loosened transplants with one hand and plop them each down into a hole. Follow through by firming the soil around each plant. Efficiently done, it takes about 90 minutes to make the holes, plant, firm and water a bed 32 feet long by 42 inches wide, holding about 1,000 plants.

Plants should never be planted into dry soil, for this immediately robs the roots of moisture and causes stress. Nor should the soil be too wet, because handling soil when it is too wet causes the soil particles to glue together, thereby reducing the flow of air, water, nutrients and microorganisms. The best way is to water the bed lightly late on the day before planting, and then to water thoroughly within 30 minutes after planting. You can be more relaxed if it is cloudy. That first watering should contain liquid seaweed, fish emulsion or compost tea.

For best results, seedlings need to be planted with their full intact root system and at the right depth, not too shallow nor too deep. Again, if you plant too shallow, the neck will flop over and you can well imagine that a neck that flops over back and forth with the wind is likely to be less efficient at transferring the nutrients than a straight, stable neck. So tuck the plants in up to the base of the leaves, but of course not covering the growing-tip. Firm the soil around the roots with a good push all around. The soil should not be too tight, but not too loose either.

The details may sound picky and time-consuming, but are important if you want the best productivity. If you train yourself to do things right, it takes no more time to do a task well than to do it poorly, and your reward will be ever so much greater.

Weeding

Very simply, remove weeds before they become large enough to interfere with the crop plants. Generally this means before they are 1 inch tall. If you let the weeds get ahead, they will rob the nutrients from the crop plants, and their roots will become so entangled with the roots of the crop plants that they will cause injury when pulled out.

Never let weeds go to seed or the weeds will come back to haunt you a hundredfold. It is easy to keep the greenhouse weed-free. But in the outside garden weed seeds come flying in from all over, and it really helps to keep the garden covered with leaf mulch. Be careful with other mulches, because some contain grass and weed seeds. Also, be sure that the compost you add does not contain weed seeds.

Watering

I have found that watering is the single most difficult task for people to understand and learn to do well. Some people want quantitative answers to questions of when and how much to water, but that is not possible to say. Some books give directions about squeezing the soil and seeing how soon it falls apart or seeing how fast the soil absorbs a puddle of water and from that deduce whether watering is needed. Both of these methods can be highly misleading because the result depends so much on the quality of the soil. It is essential to understand the principles of watering and then to learn by intuition and experience. Some people may never be able to learn how to water, because they cannot open up their intuitive powers to the needs of the plants. Perhaps nothing can ruin the salad business faster than overwatering or underwatering, and therefore it is essential that this task be done by someone who truly knows how.

To teach watering, a considerable amount of training is required, shoulder to shoulder, in silence, and with total attention on the place where the end of the water spray meets the plants and the top of the soil.

Do not water when the soil has sufficient moisture or else the soil cannot breath properly. Bore test holes here and there, a foot or more deep. The soil may look damp on the surface and 3 inches deep, but it can be bone-dry below. It is truly unbelievable how water-resistant some soils become once they really dry out and how much water is needed to remoisten them. A good mulch is really helpful for preventing such extreme dryness.

Do water thoroughly when the soil does not have sufficient moisture, and make sure that the water reaches the whole root system. Shallow watering encourages the roots to develop close to the surface, which in turn makes them more vulnerable to drying out; it also makes them unable to utilize the nutrients deeper down.

Here is an exception: sometimes the plants will wilt a bit in the heat of the sun even if the soil is sufficiently moist, and at such times it is of great benefit to give the leaves a quickie cold shower. This does not add much moisture to the soil, and it can therefore be done safely even if the soil does not need watering.

Underwatering and overwatering are both stressful to the plants, and again, any stress will cause the plants to bolt sooner than they would without the stress. Premature bolting then requires more frequent replacement of the plants, and thus requires more work. Worse yet, it can cause a shortage in production, making it impossible to meet market demand. Thus you may loose your customers to the competition.

When watering, do not direct the spray straight down onto the soil, because the water pressure will cause soil compaction. The same applies when you water onto plants because of the possibility of injuring the leaves. Instead, direct the spray upward at an angle sufficient to cause the droplets to fall by gravity rather than pump pressure.

Keep adjusting the volume control. It can be full on when watering beds at a distance, but turn it down as you water closer to where you are standing. Also, wave the spray back and forth in a rapid shivering motion, as this avoids puddling and runoff.

Once you get good at watering you can use your other hand to inspect for insects, pull some weeds, prune some plants. You will also be dreaming of various ways to automate the watering.

The best automatic watering system I know of is the C.I.T. underground ceramic irrigation system that was installed in the greenhouse, which I described earlier. Because this system drastically reduces the amount of water required for irrigation and causes the roots to go deep down away from the hot surface of the soil, I wish I had the C.I.T. system in the outside garden as well. But I received only enough pipes for the greenhouse. Instead, the outside garden is equipped with various soaker hoses and overhead oscillating sprinklers. Newly planted or sown beds are watered by hand with soft spray wands. Do water deeply to encourage deep root development. Avoid using the sprinklers on windy days, as the water will be unevenly dispersed. On hot sunny days do the watering in the early morning or late afternoon to reduce water loss due to evaporation.

Harvesting

On a sunny day the harvest must start at the crack of dawn so that it can be completed before the mid-day heat, which sets in by 11 a.m.

You can harvest leaf by leaf, cutting one to three of the outer leaves at the 4- to 5-inch stage, leaving the inner crown of smaller leaves. Three to four days later the smaller leaves will have grown to harvest size. This is the best way to get even-sized perfect leaves at their peak flavor and nutrition and also the most predictable rate of production. This is the method I have mostly used.

An important warning: do not harvest leaves too small. When there is more demand than supply, it is tempting to harvest the smaller leaves in order to get up to the requested poundage. This creates a dangerous downward spiral for a couple of reasons. One is that if the leaves are half the optimum size, twice as many are needed to make a pound, which means harvesting takes twice as long. The other reason is that the leaves that remain on the plants are smaller than usual and therefore less capable of photosynthesis, which in turn reduces the rate of production, resulting in even less poundage the following week. The plants are also stressed by such close cropping, which results in earlier bolting.

Instead, you must anticipate the possibility of a shortfall about two weeks ahead, and compensate for it by sowing the number of buckwheat trays that you need to make up the shortfall. An 11-by-21-inch tray of well developed buckwheat can yield almost half a pound, and trays can be tucked in here and there everywhere. Buckwheat tastes delicious and can make up about 20 percent of the total salad without reducing the quality of the salad mix.

Another method of harvesting, different from the leaf-by-leaf method, consists of cutting across the whole plant about one inch above the base of the plant, leaving just enough so the core can regrow. It takes about two weeks for the plant to be ready for the next cutting, and you may be able to do this three to four times with the same plant. I sometimes used this methods with some varieties such as arugula and romaine, but I do prefer the leaf-by-leaf method because it produces a better quality product, tastier, more nutritious, and with longer shelf life.

Whatever harvesting method you use, handle the leaves carefully so they do not bruise. My general rule for handling the leaves, whether for transplanting or planting, harvesting, washing or packing, is to handle them in such a way that you can imagine an 1/8-inch air space between your hand and the leaves. Also, be very careful, whether cutting the leaves with scissors or snipping with your fingers, not to injure next week's harvest. A snip or a snap will leave a brown scar that will easily spoil. In addition, be careful not to tug at the roots of the plants even slightly, for this will break off myriads of tiny primary feeding roots. It may not kill the plants, but it will cause stress and therefore reduce the productivity.

Put the leaves into a light-colored plastic tub that you drag along the rows with you, and on sunny or windy days keep the tub covered with a damp white towel to prevent the leaves from wilting. We keep the salad leaves separated into about 20 different categories: green lettuces, red lettuces, and one of each of the other 18 or so different varieties of greens and herbs.

Some people never learn to be efficient harvesters, whereas others quickly learn to be whizes, both fast and careful, harvesting 8 to 12 pounds of perfect leaves per hour.

Washing, Draining, Packing

The washing system should be built to a comfortable working height to minimize back strain, and it needs to be in the shade and, as much as possible, out of the wind. Children's plastic pools, about 1 foot deep and 5 feet in diameter, make good wash tubs, and they can be set on tables made from cable spools.

Set up the harvest tubs of different greens in a circle, like a palette, and sprinkle a portion of the contents of each tub into the pools. With a light shivering motion, carefully blend the different varieties. Again, imagine an 1/8-inch space between your fingers and the leaves. Dirt will fall off the leaves to the bottom, while lighter stuff like insects and mulch will float to the top. Skim the floaters off the surface as you carefully blend the leaves. This is also an opportunity to edit out lesser-quality leaves that were inadvertently included during the harvest.

Let the leaves soak for at least five minutes, at most 30 minutes. The colder the water, the crisper, stronger and longer-lasting the salad will be. In the summer it helps to add a block of ice to the washing pools. Then scoop up the leaves, gently, and place them in drainage hampers. Laundry hampers, about 16 inches in diameter and 30 inches tall, made of smooth plastic with many perforations, work well and are capable of holding about 10 pounds of greens.

It is important that the leaves drain off as thoroughly as possible, because puddles of water in the bags will cause them to rot quickly and will greatly shorten the shelf life. Set the hampers on their sides with lids secured, and repeatedly turn and lightly shake them. To prevent the outer leaves from wilting, keep the hampers tightly covered. A leaf can wilt beyond recovery ever so quickly if exposed to sun or wind.

A centrifuge system is even better. Some brands of washing machine can be adapted by removing the center agitator. Set the hamper in firmly and let it spin for a few seconds, just enough to eliminate the water drops. Too much spinning will crease the leaves, and any place where the leaf has been creased will be the first place to spoil.

When drained, gently pack the salad into plastic bags, with an eagle eye open during this last opportunity to edit out poor-quality leaves. I recommend 1/2-pound, lightweight Ziploc bags for distribution in stores and 2-pound bags for restaurants. Keep the bags cold until delivery.

The washing-draining-packing operation could be greatly improved to reduce processing time. I dream of a streamlined system consisting of a long trough, 8 inches deep, with water overflowing to a filter below and recirculated with a small pump back up to the trough. Mixed salad greens are sprinkled into the water at the beginning of the trough, culled as they flow "downstream" and scooped into drainage baskets. Several drainage baskets fit into a 5-foot- diameter drum that is attached to a wheel equipped with a pulley, a larger version of the starter pull on a lawn mower. Three to four pulls is all it would take to adequately dry 50 pounds of salad greens. Thus two people could wash and drain about 300 pounds per hour.

I also dream of an improved bagging system. Empty one drainage basket at a time into a large plastic funnel suspended above a conveyor that holds the plastic bags open. Right under the funnel is a scale to ensure that each bag receives the right amount. Such an operation would greatly speed the packaging process, and would also minimize handling of the leaves, thus increasing shelf life.

Marketing

When there is competition from other growers, it is important to maintain continuous marketing in order to sell every leaf of this wonderful product.

Wholesale directly to restaurants and stores rather than through dealers. This way you get the best price and have the best control over the product that the client actually receives.

To restaurants and caterers: Send an attractive flier that will fit a three-ring binder, describing in minimum number of words all pertinent details about the product. Include a great photograph of a 1-ounce salad serving, as well as photographs of the farm operation. Follow up with a phone call to make an appointment with the chef or chief buyer.

Arrive, on time, with the following: one 8-inch paper plate containing a perfect ounce of salad mix (accurately weighed on a sensitive scale), one with a perfect 3/4 ounce, and one with a perfect 1.5 ounce, each sealed into a Ziploc bag. This way the buyer will know exactly how much each serving contains and what it will cost. Help them figure out how much profit they can make per serving and point out how little work will be involved for them to produce such a superior salad plate. Also bring a 2-pound bag, so they will know how it will fit in their cooler and on the counter. Encourage tasting and handling, and discuss price and terms. Let them know that you can deliver this perfect salad once or twice a week regularly all year-round. Point out that there will be no waste, except the plastic bags (I dream of a reusable container). Leave the salads for them to test-use and also leave additional information, with names of all the varieties. Invite them to visit the farm.

After they become customers, be unfailingly faithful with quality and delivery, and keep in close touch to check if all is satifactory. Restaurants are high-stress workplaces, and chefs put a high value on absolute dependability. From time to time, add special little touches, such as new varieties, bunches of special herbs, and edible flowers (these can also become a separate profit center). Some clients will want poundage of separate varieties, such as arugula or radicchio. Try to accommodate, but this is often more complex to manage, so make sure it makes economic sense to you.

Stores: In order to minimize price markup, you must offer to take back or give credit for any product not sold. Help the store owner figure out what the profit per square foot can be with this product. We have long been getting a wholesale price of $5.25 per 1/2-pound bag, which retails for $6.99, a 33 percent markup. About 12 bags can fit standing up in a space 15 inches wide and 24 inches front to back, and it requires very little of the produce manager's time to keep this space stocked from boxes in the back cooler. A store selling 200 bags of salad per week would thus earn a gross profit of $348 per week from 2.5 square feet, or $7,238 per square foot per year, without any fuss or waste.

Make an attractive and informative sign for the front of the space, well protected from being faded, bent or lost. At a retail price of $6.99 for a 1/2-pound bag, a 1-ounce serving will cost 87 cents, and a whole bag will provide salad for an elegant dinner for 8 to 12. What an easy and delicious value!

The best profit can be gained by selling direct from the farm. I highly recommend that this opportunity be optimized if your location makes it possible. This can be done by combining a number of different attractions: a cafe/store with a salad bar, tours, and of course visiting the animals. Selling gorgeous 1-ounce salad servings for $2.50 brings the gross income up to over $30 per pound. Adding a limited menu that incorporates organic eggs, lamb, chicken and rabbit, plus a few organic home-grown herb teas, fresh-pressed organic juices (the pulp becomes prime animal food), plus some yummy desserts, and you've got yourself a nice little farm business beyond your wildest dreams that also happens to be immensely beneficial for other people and for Earth.

❧ ❧ ❧

A YEAR AT SOLVIVA

Scheduling will depend on your location. These recommendations are meant for a climate similar to Martha's Vineyard's where the average last frost is April 25 and the average first frost October 25.

DECEMBER and JANUARY

During these months the sun rises after 7 and sets about 4:30. The sun is low in the sky and is often concealed by thick clouds for days on end. These are the two months with the lowest productivity. For about two weeks around the turn of the year it seems that growth almost stops. It soon picks up again, and by taking certain measures high yields can be maintained. Follow the recommendations for temperature control and watering.

About every two weeks sow enough seeds to replace whatever varieties may be exhausted some five to six weeks later. (It takes about four to six weeks to raise a seed to mature transplant at this time of year.) If you end up with more young transplants than you have space to plant, you won't have any trouble selling them in six-packs or window boxes to people with sunny spaces (yet another strong profit center potential). Or you can grow them out as big as they will get in the seed trays or six-packs, and then clear-cut them and add to the harvest. If you plant too few seeds, you will have a shortage of product down the line.

Grow buckwheat to increase the poundage. Put about an inch of growtube mix into seeding trays, sow densely spaced buckwheat seeds, water with seaweed solution, and tuck in as many of these trays as you can fit in among the seedling trays and growtubes upstairs. It only takes about two weeks to grow to beautiful 2-inch heart-shaped leaves. Each tray can yield about half a pound, which takes less than a minute to harvest. After clear-cutting, with 3-inch stems, dump the trays into the chicken room.

This is a good time to take a vacation for a week or even a month, provided of course that someone totally dependable can remain in charge of watering, venting and harvesting, and caring for the animals.

FEBRUARY

The light is returning, and the plants are growing faster. The sunlight is becoming more powerful, so be careful to avoid overheating. Keep on sowing seeds every two weeks, buckwheat weekly. Since this is still a slow period, it is a good time to do maintenance and repairs and to lay in a good stock of supplies such as plastic bags, boxes and labels.

Check the compost bins to make sure the breakdown is proceeding properly. If not, turn the compost to speed up the process.

If you see any sign of rats, take appropriate measures immediately, or else they can become a terrible problem.

Order all the seeds you will need for the next six months, including new varieties that sound promising. When the seeds arrive, place them in Ziploc bags, totally closed so no humidity gets in. Organize the seeds into several different categories, such as green lettuces, red lettuces, stronger-flavored greens, herbs, edible flowers.

MARCH

The production is booming and the cycles of seeding and bolting continue. Make sure you have on hand plenty of grow-mix, compost, trays and labels, because soon it is time to drastically increase the sowing for the first wave of planting the outdoor garden.

In the last week of March sow into seeding trays about 12,000 seeds per 1,000 square feet of actual bed space for the first planting of the outdoor garden. To ensure even production during the summer, rather than a giant boom in June when everyone else's garden is also booming, I recommend sowing seeds for only about a quarter of the outside garden in this first wave.

During March you can expect an increase in the insect population, so order what you need: ladybugs and lacewings for sure, and perhaps also predatory mites and Encarsia formosa. Make sure they are shipped in safe weather and release them according to the directions. I recommend ordering a half-gallon of ladybugs, which is about 25,000 (only about $25!). Release about a quarter of them right away and put the rest in the refrigerator where they go dormant. Moisten the bag every few days, and they can stay alive for several months. Release more ladybugs in batches every two to three weeks thereafter, or as you see a need.

APRIL

Sow another 12,000 seeds per 1,000 square feet of growing bed for the second wave, about two weeks after the first one. Remember that if you sow too much, you can clear-cut the seedlings and add them to the salad mix, or you can sell them in six-packs.

Prepare the outdoor garden as described earlier. For a garden that has in a previous year already been prepared with rototilling, building the beds and covering with shredded leaves, all that has to be done now to prepare for sowing and planting is to rake the leaves off the tops of the beds.

Make sure that all hoses, sprinklers, tools, carts and wheelbarrows are in good repair.

Sometimes the weather is warm enough for the first planting of salad greens to take place the last week of April. But there is really no advantage in pushing it, because they won't be harvestable any sooner than if you plant them in early May. Of course you can plant peas earlier, but in this book I am primarily dealing with salad greens.

Make sure there is no way for racoons or other pest animals to enter the greenhouse or the animal areas. They become highly motivated to find food and shelter when it gets close to giving birth, and they are smarter at getting through doors and vents than you can possibly imagine.

MAY

By early May the soil will be ready for both planting and sowing salad greens. The growth is painfully slow the entire month of May. The plants just sit there without growing for weeks on end. You will think that the soil is poor, or that you've lost your green thumb, but actually down below the soil surface the roots are burrowing deep and laying a powerful foundation for strong growth. Be patient.

The greenhouse, by contrast, is booming. There will be some very hot days which will prompt some of the older plants to bolt. Clear-cut those plants that are about to bolt, before they turn bitter.

The insects in the greenhouse are also booming. Don't be tempted to resort to toxic measures, but do keep up with the harmonious methods and trust that harmony will prevail.

May is a bridge month between the inside and the outside production. The trick is to keep the greenhouse in high productivity until the garden kicks in. Keep those buckwheat trays coming.

As soon as night temperatures no longer dip below 50 degrees F all the greenhouse vents can remain fully open until cold weather returns in the fall.

JUNE

It always seems that on June first the outside garden suddenly starts to explode. Because the cooler May has been great for root growth, June is more productive for salad greens than any other month.

June is also the time when everyone else's garden is booming, and most growers have more product than they can sell. Some chefs and grocers will be sorely tempted to decrease their orders from you because at this time they can buy salad ingredients for less from others. Try to retain your clients by special treats and freebies, and point out to them that you will be faithful to them in the crunch month of August and all the rest of the year if they will stay faithful to you during the month of June. You may need to lower the price a bit until the early summer flush has passed.

JULY

This is often a month with little or no rain and sweltering heat. Proper irrigation is essential, as is continuing seeding and planting, soil enrichment and weeding. Mulching is highly beneficial to keep the roots cooler. Demand on the Vineyard and many other summer resort places is approaching the highest point. Don't forget to take time off, or you will burn out.

AUGUST

This month is even more likely to be dry and hot. By this time most other growers have little or no salad greens. But you have been doing what needs to be done and will therefore be able to fulfill the peak demand.

Keep seed trays coming for the outside garden, because with a little protection outdoor production can often continue until Halloween or longer.

SEPTEMBER

Early in the month it is time to begin to prepare for the fall production in the greenhouse and the cold frames. Sow enough to replace whatever plants will be bolting by early October, plus 25 percent more. Extras can be sold in six-packs.

OCTOBER

This is another bridge month, like May. The garden can still be kept highly productive, and the fall production in the greenhouse is just beginning.

Cover the outside garden with protection against frost. It is a nuisance but well worth it here, for with proper precautions the production can be bountiful until Thanksgiving. If there is a chance of frost, water the garden in the afternoon so the leaves will be wet during the night. This helps to protect the leaves from damage by frost. Sometimes the whole garden will be white with frost in the morning. If you do not touch the leaves until they have defrosted, you will find that they will fully recover, unless of course it was a really hard frost.

As soon as night temperatures begin to fall below about 45 degrees, it is time to close the greenhouse vents. However, great care has to be taken to open them up fully on any sunny day.

NOVEMBER

Because of the long fall here, November is still a bridge month, but you cannot count on it. Any night now the temperature can plummet so low that movable covers can no longer protect the plants from death. Soon the garden soil will be hard as rock, but inside good cold frames hardy plants can survive and be productive even through severe cold.

Before the snow descends, get more chipped leaves, enough to cover the whole garden to a depth of 2 to 3 inches. This blanket, plus the winter's periodic blankets of snows, will protect the earthworms, pill bugs, millipedes, earwigs and the billions of wondrous little microscopic creatures that inhabit this rich soil, keeping them happy and active, patiently awaiting the coming of spring and yet another abundant growing season.

ANIMALS

The animals make a charming and lovable addition to the Solviva farm, and they greatly boost the yield of salad greens and other crops. But because there are those who believe that it is inherently cruel to keep animals, and crueler yet to kill them, I want first of all to address the very valid concerns that surround these issues. Most conventional large-scale factory methods of raising chickens, rabbits, cows and other animals for the production of eggs, milk and meat are unspeakably inhumane, polluting and wasteful. I will not go into the gruesome details about what happens to cows given hormones to maximize milk production, or to chickens who live their entire lives crammed four to five per small wire cage, or to sheep who await slaughter strung up by one leg on overhead conveyors.

Many who find out the truth become vegetarians as a result. Others give up meat because of the many harmful substances that are contained in most industrial meat. Not only are there dozens of different chemicals that the animals are subjected to for speeding up their weight gain, increasing milk and egg production, and controlling pests and diseases, but there are also the harmful substances that the animals themselves produce as a result of the stress they undergo while living and dying. I myself was a vegetarian for several years, before I started raising my own farm animals.

Now I know that it is possible to keep animals in ways that provide them a happy and healthy life. I also know that it is possible to kill animals in ways that are free of pain and fear. In my opinion, these are the basic requirements for keeping or eating animals. I believe we can eat meat, in moderation, without causing suffering or harm. I will go one step further: I believe that we need animals in order to enrich our soil and our diet. Even if we do not eat the meat, the vegetables we eat will be more nutritious if they are grown with compost that contains animal manure.

Our emotions tend to run wild, our minds tend to close when confronted with the issue of death and dying. Why is it that we are so revolted by the idea of being in control of the moment of death? Why would we rather keep a human being alive as long as possible with artificial means, against her will, perhaps spending hundreds of thousands of dollars in the process, rather than helping her to die peacefully? Some people get so riled up on either side of the fence on these issues that they are ready to kill those who disagree with them. I believe that there is nothing bad per se about death. Death is an absolutely essential part of life. Life cannot exist without death. We will all die some day, it is only a matter of how and when.

Like most of us I have little trouble killing a mosquito or a tick. But could you kill a chicken and feel okay about it? Well, I did, once, and I could do it again. I feel certain that, because of the

Photo by Alison Shaw

Anna with Jenny Donkey, Heather and her little lamb.

way I prepared her and killed her, this chicken died without any fear or pain. She struggled when I first picked her up, as chickens usually do, but as I held her securely in my arms, stroked her neck gently and sang to her softly, she relaxed completely within a few seconds. In that state of relaxation she was entirely free of fear. Then I laid her head on a stable block of wood, and, continuing all along to stroke her neck and sing, I held a sharp axe just over her neck and then came down hard on the axe with a hammer. Her head was severed in one blow, and her body never did go into the spasms that people associate with a chicken being killed, but remained completely relaxed. It was very difficult for me. In fact I held her in my arms in that relaxed state for more than half an hour before I finally could get myself to do the deed. But I am certain that she experienced neither fear nor pain.

But could I bring myself to kill a rabbit or a sheep? Even though I know from many experiences that even the most terrified rabbit or sheep can be completely calmed within a few seconds with the same method of holding securely and singing and talking gently, and that a sharp knife inserted in the right place would bring instant and painfree death, I would suffer terror. I clearly recognize that I too have trouble dealing with death.

I consider animals to be an essential part of food production, as essential as yin is to yang. Their manure makes the best compost fertilizer, whereas compost without animal manure is anemic by comparison. Animals provide food, fiber, feathers - and love. Many people can testify to the love contact that is possible with a dog or cat or horse. The same is possible with sheep and rabbits, and, yes, even chickens.

Animals can graze on land not suitable for growing grain or vegetables for people, without causing soil erosion. On other land, food plants can be grown that cannot be digested by people but can be digested by animals, who in turn can be digested by people.

I am totally opposed to producing meat in the industrial ways that are prevalent today all over the world. These methods cause devastating soil erosion and pollution and depletion of water resources. They consume enormous amounts of fossil fuels and destroy rainforests. They cause homelessness, disease, suffering and death among human beings, and even extinction of species and whole ecosystems.

I disagree with those who believe that beef, or any other meat, is inherently bad food. I believe meat is an excellent food and that it can be produced in friendly ways, provided that meat consumption is greatly reduced below current levels.

There are people who believe that meat production requires many times more land and water than the equivalent amount of protein production in the form of plants. I disagree. I believe it is possible to produce more protein per square foot of land and gallon of water by wisely combining plants and animals. I also wonder about the sincerity of those who moralize against killing animals, yet wear leather shoes, carry leather bags, eat yogurt and ice cream, milk and cheese, and use blood meal to fertilize their gardens.

The 10,000-square-foot Solviva greenhouse that is described in the projected one-acre farm is designed to comfortably accommodate some 400 chickens and 100 rabbits, with plenty of space for all to roam around. The rabbits can be grown to market size in less than five months. Half of the chickens can be marketed at 7 to 12 weeks, as fryers or roasters, while the rest of the chickens can be marketed after having been productive egg layers for a year. Thus, one acre of land can produce 40,000 organic eggs per year, plus over 7,000 pounds of organic chicken and rabbit, in addition to the 50,000 pounds of salad greens previously mentioned. Some of the food for the chickens can be provided right on the one acre in the form of sprouts, plant wastes and weeds, trapped flies, moths, slugs, pill bugs and beetles, as well as earthworms grown in the compost bins.

The manure compost and the CO_2 from the breath and the bedding of the animals are the greatest reasons for the exceptionally high yields of the plants in the Solviva greenhouse. The animals also provide heat to keep the greenhouse productive through even the worst cold snaps. Since each animal produces about 8 BTUs per hour per pound of body weight, and the average weight of these young and mature animals is around 4.5 pounds, 432,000 BTUs per 24 hours would be produced from the 400 chickens and 100 rabbits recommended for the 10,000-square-foot Solviva greenhouse. This heat could be provided by oil, gas, coal, or wood, but why suffer the expense, pollution, depletion, and insecurity (and phenomenal amount of work in the case of wood and coal) when you can get the heat free from animals? And you certainly do not have to kill the animals in order to integrate them into the greenhouse, for they are just as valuable in old age from the standpoint of heat, CO_2, and compost, and after death their bodies transform into valuable compost.

The Solviva designs and methods provide spacious, clean, comfortable and secure quarters for the animals. They are separated from the plant room by means of a waterwall, which absorbs and stores solar heat from the south and animal warmth from the north, and slowly releases the warmth as needed.

Even though sheep are not suitable for inclusion within the insulated part of the greenhouse - because they like cooler conditions - I still recommend raising a few sheep because they are so very lovable.

Rafe Brown

THE CHICKENS

In the 3,000-square-foot first Solviva greenhouse, 100 chickens live happily in a 200-square-foot section in the northeast corner of the greenhouse. Their whole floor area is deep with leaves and sawdust (not hay or straw because it forms a felted mat difficult to aerate), which together with the chicken droppings form a health-promoting compost bedding. This bedding provides luxurious dust baths which is one of the reasons why the Solviva chickens have been free of diseases or pests (such as lice and mites) ever since the beginning in 1983. The bedding also provides worms, bugs, seeds and endless entertainment to enrich the chickens' diet and quality of life. Visitors familiar with chickens comment on the exceptional health, vigor and size of the Solviva chickens.

Once every couple of weeks 3 to 4 inches of leaves or sawdust are spread over the chicken bedding. This adds carbon molecules which bind with the nitrogen molecules in the chicken droppings, and thus prevents most of the nitrogen from evaporating in the form of ammonia. Too much ammonia in the air is harmful to the chickens and can be deadly to the plants. A pitchfork is plunged into every square foot of the bedding once a week and rocked in order to add oxygen to keep the bedding aerobic. There is no need to do the much harder work of turning over the bedding, because once it is loosened by the pitchfork, the chickens will get right in there and effectively do the turning. This management system keeps the area sweet-smelling and healthy. If the bedding is allowed to become too compacted, it goes anaerobic, which produces foul odors and prevents composting. (An alternative is to purposely process the compost anaerobically in vats outside the greenhouse, and thereby produce methane gas for heating, cooking and electricity, a fascinating technology with which I have no direct experience).

The bedding is a great source of CO_2, as it is filled with billions of breathing microscopic animals. Please refer back to the section about the Earthlung filter which shows how to best transfer the CO_2 into the plant area. The bedding also offers great heat potential, which can best be utilized by setting up several compost bins made from 4-by-4-foot pallets with closely spaced boards or chicken wire on the sides right in the chicken room. Starting in early winter, scoop part of the bedding into one of the bins, filling it to the top. Water it, and within a day or two the temperature will rise to 150 degrees F or more and will stay hot for quite a while. Each time you hear of another cold snap coming, fill another bin. By the end of winter the whole floor area will have been turned into the compost bins. Keep putting in the weekly loads of leaves to cover the whole floor.

With this management, the 400 chickens that can be happily accommodated in the 1,000-square-foot chicken room in the 10,000-square-foot greenhouse will yield over 1,000 cubic feet of superb, weed-free compost. Chicken compost is pretty strong stuff, so I like to let it mellow in the bins for a couple of months. After that, it is ready to be transformed into nutritious, delicious Solviva Salad. Pretty amazing when you think about it.

The chickens like to roost aboveground for the night. There needs to be almost one foot of roosting space per bird, plus about 25 percent extra to allow for the shakeout of their social life. The roosts (all set at the same level so they don't squabble about who gets the highest perch) are built as frames that are hinged to the wall. Thus they can be raised out of the way to allow easy maintenance of the bedding under the roosting areas. The roosting frames need to be very strong in order not to break under the squirming weight of the birds. A 4-by-8-foot frame will hold 30 to 35 chickens, which together can weigh over 250 pounds. The roosting supports can be made of tree branches about 2 inches in diameter. These are comfortable and secure for the chickens to grasp, and, having a rounded top surface, the droppings are more likely to fall to the ground instead of remaining on the roost.

Laying hens need nest boxes, about 10 per 100 hens, about 12 by 12 by 12 inches and 24 inches off the floor. (This way they are still off the floor when the bedding builds up to 18 inches). Keep a barrel of shavings close to the nest boxes, and sprinkle in some shavings in each nest right after the daily egg gathering. The chickens will poop in their nests and the poop does stick to the eggs, so this daily covering saves enormous amounts of time in egg washing. Once a week, scoop out the entire contents of each nest, right onto the bedding below, and line the nests with fresh shavings. Keeping the nest boxes clean and harvesting the eggs daily is necessary to prevent the terrible nuisance and waste of the chickens pecking, breaking and eating their own eggs.

Because these chickens breathe clean air, are free-ranging, and eat fresh greens and earthworms year-round, their eggs are said to be 25 percent lower in calories and cholesterol than ordinary eggs, as well as higher in protein and vitamins. It is easy to understand why there is a major difference in quality when you learn that factory hens are bunched up so close in cages that they can hardly turn around, and the air they breathe is so foul with ammonia that workers have to wear gas masks.

Over the years we have killed a few old hens in the humane way described earlier, but we have not regularly "harvested" them. Thus, some of the chickens have lived to be seven, eight or nine years old, maybe more. You can tell by their eyes, combs, wattles and legs that they are exceptionally healthy. They have never had any of the prevalent chicken pests or diseases. I feel that the major reason for this is the deep bedding system and the healthy ecosystem that develops in it. Some of the chickens have died from old age, some have been killed by hawks, and there have been tragic times when raccoons have broken in and killed several chickens per night before we could finally figure out how to stop them effectively. Raccoons are very smart and very brutal (from my point of view), but they, and the almost equally destructive skunks and rats, can be prevented by a pestproof foundation, which is of course now integrated into the latest upgraded Solviva greenhouse and home designs.

I recommend the varieties of chickens that I have raised, Rhode Island Reds and Plymouth Rocks, or hybrids close to these. They are peaceful, handsome and intelligent, and each lay about 200 eggs per year. You could raise super-duper egg-laying machines like Leghorns, capable of laying over 250 eggs per year, but I hear that they are neurotic and cannibalistic. Or you could raise Araucanas that lay small colored Easter eggs. Or you could raise any number of different fancy breeds with extraordinary (and impractical) plumes on their heads and tufts on their feet. If you enjoy almost dying with laughter, get a catalogue from one of the places that supply fancy chickens. It is interesting to consider the fact that these chicken varieties, along with all modern breeds of farm animals, and dogs and cats, have been evolved by man over the centuries.

Every couple of years we replenish the chickens by ordering day-old chicks from a reputable poultry supplier. We generally order 50 pullets at a time, and usually at least one of those will turn out to be a rooster. We get an urgent call from the post office ("Come down quick!"), and we cause some smiles as we pick up this loudly peeping box.

Within the chicken coop is a separate space for raising the chicks until they are old enough to fend for themselves among the adults. It is enclosed with 1-inch chicken wire, so they can get acquainted by sight. As soon as we arrive home with the new chicks, we open up one corner of the box and carefully remove one chick at a time. (Why is it that baby chicks, ducks, geese and other poultry birds are so adorable right from hatching, whereas most other birds are rather ugly?) First of all, they need to learn to drink, so we dip the beak of each one into the water container. They require cozy warmth, and this is accomplished by suspending an upside-down cardboard box just above the bedding. In this box hangs a 60-watt light bulb in a wide reflector shade. The box has a little door cut out on each side, so the chicks can go in and out. If we find the chicks standing on tiptoe crowded under the light bulb, it means they are not warm enough, and the solution to that is either to lower the bulb or increase the wattage. If, on the other hand, the space under the light bulb is vacant, the chicks are too hot, so we raise the light bulb and box or decrease the wattage. If the chicks are evenly distributed, all is well. They are so smart.

Usually there is at least one hen who is yearning to get in there to mother the chicks, pacing and purring back and forth along the chicken wire enclosure. At first I thought the hens had evil intent, but then under my watchful eyes I let one in to the chicks. What a lovely scene followed. She immediately went to the little chicks, and as she cooed and purred in the most comforting, motherly way, the chicks gathered in under her wings. Soon she was the biggest, fluffiest hen you ever saw, with little chick heads protruding everywhere. Every one of the 50 little chicks had managed to fit in under her. Of all the happy hens I have seen here over the years, she was the most. Ah, the joy of mothering...

After a few weeks the chicks have grown tremendously, have lost their down and are dressed in feathers. They have become rangy teenagers and are ready to be let out with the rest of the flock, as they are then old enough to run away from any hen who may chase them with cannibalistic intent.

At Solviva there has never been trouble with cannibalism, nor have new chicks ever picked up diseases. This is in spite of the fact that in all these years we have never followed the conventional practice of cleaning out and sanitizing the chick quarters before the new chicks arrive. This is yet another proof that Solviva's deep-bedding system works to the benefit of all.

The first rooster I had turned out to be a real meany. He constantly brutalized the ladies (or so it seemed to me), and he was truly dangerous to people. He would come charging around the corner to attack us with his formidable talons and sharp beak, the very picture of explosive fury (why??). A brave intern with farm experience initiated the chopping block with this rooster, and we all feasted. I never understood why this rooster turned into such a brute, nor why none of the subsequent roosters ever did. All the other Solviva roosters have been perfect gentlemen both with each other and with their ladies, as well as with us.

You can probably tell that I love the chickens. Contrary to popular belief, they are quiet and clean. Because of the way they are managed, it never becomes smelly or messy inside or out. The birds are friendly, curious and very intelligent. I like just to sit in their coop, singing and crooning with them about this and that - chicken opera. I can pick one up, and as soon as I stroke her neck and head, she'll be instantly relaxed and in bliss.

The chicken room is equipped with a small trapdoor that opens up to become an exit ramp (set high enough to allow for the buildup of the bedding). Since the hens lay their eggs in the morning, the door stays closed until about 1 p.m. This way we don't have to search for eggs in nooks and crannies all over the farm. They all come streaming out, clucking and crowing (for just one minute a day they are rather noisy), joyfully pecking for worms, seeds, sand and other goodies. Over the years their yard has become as fine and green as a golf course, self-established with some wonderful wild species of grass that manage to thrive in spite of their pecking and scratching. This is one of the many signs of harmonious management.

The chickens also roam widely in among the sheep on the larger pastures, but they are not allowed in among the gardens. This is because in the process of seeking their favorite food, bugs and worms, they scratch the ground - and the small salad plants - with their powerful feet. A combination of chicken wire that extends 6 inches below ground to 4 feet above ground, and two strands of wire to a height of 5 feet, plus trimming the feathers on one wing once a year, prevents them from getting under or over the fence. It is important to trim just one wing as this greatly reduces their ability to fly. The process of trimming a wing is totally painless, like trimming your fingernails, and it gives you a chance to have a brief one-on-one with each chicken, which is highly beneficial for both you and them.

The chickens need a constant supply of clean water, which can be easily accomplished with a float valve attached to the inside of a small pan. They also have a constant supply of grain that comes down a chute from a hopper as they eat from the attached bowl at the base. The reliable supply of food keeps them relaxed and secure, and there has never been any problem with overeating.

We built the grain hopper up against the east exterior wall, and it is ratproof. The delivery truck can back right up to the hopper, and the truck bed is at the same height as the hopper platform. The driver just props the hopper lid open and lays in eight 100-pound bags, four on each side of the center hole. One bag at a time is opened, and the grain slides down to fill the chute. Whenever possible let gravity do the work. For the larger 10,000-square-foot greenhouse operation I recommend buying a standard grain hopper for bulk delivery of a season's worth of grain.

A hundred chickens take no more than 10 minutes a day of care, including collecting and boxing an average of four dozen eggs and spreading fresh shavings on the 10 nests. It takes a lot longer if the nests are not kept clean, or if you don't collect the eggs daily, because you then have to soak the eggs and scrub off the chicken droppings. Once a week it takes a few extra minutes to add more leaves or sawdust, to aerate the bedding, and to scoop out and replace the nest linings. They will stop laying as the days grow shorter, but this can be prevented by suspending a low-watt light in the coop, with a timer to turn it on just before dusk and off a few hours later to provide them with a total of 14 hours of light.

The chickens more than pay for the cost of their feed with the eggs they lay. The compost, CO_2 and heat are invaluable fringe benefits.

The Rabbits

Even though I no longer have the rabbits, I still recommend them if you are willing to conscientiously keep up with the required daily care. Thirty rabbits require about 15 minutes a day, plus an extra 20 minutes every week or two for a deeper cleaning. This is really not much more time than the chickens, but far more work and more problems if you don't keep it up properly (a stitch in time saves 99 in this case). In addition, if you have angora rabbits, about 20 minutes per rabbit per month is required for harvesting the angora fiber. I had to give away the Solviva rabbits several years ago, because when I was no longer managing the place, the daily care was not properly maintained and as a result, the rabbits became unhappy and unhealthy.

Mine were gorgeous angora rabbits in gold and silver, black, fawn, gray-brown and white. I chose angora rabbits because I wanted to garner the cloud-soft fiber and mix it with the sheep wool to make an exceptionally soft and warm angora/wool yarn. Every month or so I groomed a magnificent 1-ounce angora cloud from each rabbit. By holding securely, singing and talking softly and gently pulling small tufts of angora, this grooming can actually be a pleasurable experience for the rabbit.

I send the sheep wool and the angora fleece up to Green Mountain Spinnery, a small custom spinnery in Putney, Vermont. They were the first to provide custom spinning. The yarn ends up containing 6 percent angora and 94 percent wool, which I feel are perfect proportions. Then I dye this soft yarn in my studio and end up with vibrant, subtle colors. The yarn is sold at premium price direct to customers. A flock of 25 sheep and 30 angora rabbits, plus some wool from select Vermont sheep, can produce 200 pounds of the softest wool you can imagine, bringing in over $12,000 in gross income from the dyed yarn, or $30,000 if the yarn is transformed into sweaters and blankets. I still produce this beautiful yarn, but nowadays, because I no longer have the rabbits, I have to buy the angora fiber from another producer. Even though it is extremely costly, it is worth every penny because of its remarkable ability to make wool softer, lighter and warmer.

I strongly disagree with the standard recommendations for rabbit management: one rabbit per cage, except for mothers with babies, wire mesh floors without any bedding, and nothing in the cage except a bowl of pellets and a water bottle. I have also heard it recommended that rabbits not be given greens ("it causes diarrhea"). I don't understand how such strange ideas evolve, so contrary to Nature, and why people are willing to follow them.

No wonder domestic rabbits have the reputation for being neurotic and prone to diseases, and no wonder they get infections on their feet and hind legs (hocks). This conventional way of keeping rabbits is cruel. All we need to do to begin to understand what is best for any animal, or even any plant, is just to imagine ourselves in their place. I imagined myself as a rabbit and came up with what I feel are far better ways of caring for them.

I recommend letting the does and the young live together in large warrens, each with a second, even a third and fourth level that they can get to. This way each rabbit has lots of room to roam and run, and each can choose whether to cuddle with the pack or rest alone in an upper corner. They love to live together in large groups of 10 to 15. They often pile on and around each other, languidly and lovingly grooming a fluffy body part that may or may not be their own.

The bucks, on the other hand, tend to be extremely territorial and aggressive toward each other. You would never guess when looking at soft, quiet rabbits that they would be capable of lightning speed and ear-piercing screams, ferocious kicking, scratching and biting. Thus most of the bucks have to live in solitary dens from the time of sexual maturity. You are lucky if you can find two bucks who can live happily together as friends, and no animal can really live happily alone. And you can't let a buck live with the does, or you end up with umpteen rabbits in no time - as I learned the hard way.

So what is to be done with the bucks after sexual maturity??? Obviously it is not possible to provide each one with the same large-size accommodations that the groups of does enjoy, so the bucks are limited to much smaller cages. Although I did not practice what I recommend in this sentence, I really do believe that the most humane thing to do with most of the bucks is to kill them, without pain or fear, of course, and then to let their physical entity be reincarnated either by eating or composting them. Keep only the healthiest two to three bucks with the best angora fleece for breeding, hoping that they can also live together as friends.

The only time that I separated a doe to live in a single cage was when I wanted her to breed. When I wanted more rabbits, I picked the finest doe who had not been pregnant for several months and the finest buck, and let them live in bliss for no more than three weeks. This length of time is specific because she is likely to get pregnant within one minute after the buck is let in, and the gestation period is 31 days. So after three weeks I removed the buck and let the doe live in peace through gestation, birth and nursing. The few times that I did not remove buck before the doe gave birth, he killed the babies.

After the buck has been removed, provide the doe with plenty of soft straw. About 24 hours before giving birth she will start building the most wonderful deep, secure nest. She then grooms the softest fur from her belly and rump, which need to be cleared anyway to allow for a sanitary birth and nursing. (Nature/Life is so intelligent...) She combs it off with her teeth and repeatedly accumulates a big wad in her cheeks, then spits it out as a soft ribbon, and lines the nest completely.

Then she gives birth to an average of seven to eight naked, blind, thumb-sized little babies. The doe spends hardly any time taking care of them. I have never understood how she can manage to feed and clean eight little babies in the 15 seconds four to five times a day that she tends them. She has an extraordinary sense of hygiene, because even though she gave birth in the nest, and the little babies live in it for more than a week, the angora lining is still so clean after the babies leave that with just a little bit of fluffing it can often be used for making the yarn. And in those short 15 seconds the babies manage to get enough of her extraordinarily rich milk to grow strong and plump enough in just a week to begin to venture out of the nest. They are perhaps the cutest among all babies of the animal kingdom. When newborn they move uncannily like miniature human newborns and make the same tiny mewing sounds. I know intimately about this because I once tried to keep an orphaned newborn bunny alive, keeping it warm tucked inside my shirt, feeding it warm milk with an eye dropper. A mutual bond soon developed, and I was heartbroken when it died a few days later.

For the floor of the warrens and cages I recommend solid sheet metal, not wire mesh. Cover the floor with a layer of "biocarbon" material: leaves and sawdust plus a small amount of compost, about 3 to 4 inches thick. Every week or two remove the bedding and put it into the compost bin. I devised a special apron that attaches to the cage along the door sill, and with a scraper in each hand it takes but 10 to 15 seconds to scrape the bedding into the apron and turn around and dump it into the compost bin. Multiply that time by 10 floor sections, and it takes just a few minutes to remove all the bedding. Then it takes another few minutes to put in fresh bedding. The spaces stay sweet and clean without any washing, because they are "dry-cleaned" by the millions of beneficial microorganisms who live in the biocarbon bedding.

In one corner of each warren, in a spot that the rabbits themselves have chosen as their toilet corner, place a box (the vegetable drawer from an old refrigerator is perfect, and free) and line it with the same biocarbon material as the bedding. This is where they will put all of their droppings and most of their urine. Every day sprinkle fresh biocarbon material into the box to cover the day's deposits. This immediately eliminates any odors. When the box is full, after a week or so, empty the whole layer cake into the compost and reline the box with fresh biocarbon mix.

This sanitation is so easy and quick, and it is so good for the rabbits. They find it ever so much nicer to hang out on soft bedding rather than wire mesh. Also, it is far more interesting for them because they can dig and burrow around in it. Perhaps most important is the fact that this deep bedding system, with its well-balanced ecosystem, keeps the rabbits exceptionally healthy. When properly maintained, we had none of the usual leg sores and diseases. Only when the harmonious management system was not properly maintained were there problems, and that is why I finally, regretfully, gave away the rabbits.

Rabbits thrive on alfalfa, especially in pelleted form. Alfalfa enriches the soil wherever it grows, because, like all leguminous plants, it absorbs nitrogen from the air (our air is 79 percent nitrogen!) and stores it in its root system. The rabbits also love salad scraps and weeds.

Rabbits are among the most efficient of all animals at converting plants to meat. The meat is delicious, much like white chicken, and it is the lowest in fat and cholesterol and most easily digested of all meats. And I am convinced that they, like other animals, can be killed without fear or pain.

I am told that you can make quite a good income from raising rabbits and selling them just before sexual maturity. This way you can avoid the problem of fighting among the bucks and keep them all together in a large walk-in den. A French movie, "Jean de Florette", showed a great example of farm rabbits being kept in large ground-level pens outdoors in conditions as close to Nature as possible.

Their manure breaks down very quickly and creates high temperatures, efficiently killing weed seeds in the process. Fly problems can be avoided in rabbit compost, and any other compost, by incorporating fly parasites, and if any flies do develop they can be trapped in various effective traps and then fed to the chickens.

Comparing compost from rabbits, chickens or sheep, rabbit is my favorite. It also seems to produce the highest population of earthworms, yet another potential profit center.

THE SHEEP

In 1981 I bought three beautiful sheep, a dark brown mama, Heather, who was a mix of Corriedale, Karakul and Finn, and her little black lamb, Honeydew, and from another farm a purebred Romney white ram lamb, Rasmus.

When I planned and built the greenhouse in 1983, the sheep were to be an integral part of the greenhouse, helping to keep it warm. But the heat contribution from the sheep did not turn out to be quite as large as I had thought. The reason is that, even though I built this very nice barn for the sheep up against the north side of the greenhouse, they still prefer to spend even cold nights out under the stars. However, on a really cold night the sheep can be shut into the barn, and then it sure is cozy in there. It is remarkable what an effective heating system a group of animals make. After the rabbits were given away, we sometimes put sheep into the rabbit room in severe cold. I finally understood the expression, "It was a three-dog night". Well, we had three- or even five-sheep nights. A punster friend remarked: "That gives a new meaning to the term BTU".

Of course, at lambing time, which occurs primarily in January and February, the barn is in much use. At such times the sheep barn is a most wondrous and lovable place to hang out. The floor is warm from the composting deep bedding, managed the same way as the chickens', with weekly additions of both carbon (leaves and sawdust) and oxygen (opening the bedding with a pitchfork). The first year I used straw and hay instead of leaves and sawdust. This proved to be disastrous because this stringy material formed a densely matted mass that was impossible to aerate. Digging this bedding out after the first winter was by far the most unpleasant task we have ever done at Solviva.

I highly recommend sheep for landscape maintenance. The pasture areas would be deep in poison ivy and brambles if it were not for the sheep, unless I continued the costly and gasoline-consuming mowing that was done before I got the sheep. The property is fenced around the perimeter and also around the gardens, the greenhouse, the beautiful wild pond, and the house. This way the sheep keep most of the property immaculately groomed, right up to the tree trunks and fences.

I could tell so many stories about the sheep to illustrate their remarkable and intelligent qualities, but in this book I will limit it to one. It was late spring and time for the annual shearing. The shearer was coming the following day, so I prepared by bringing in the usual equipment that the shearer required: plywood to be used as a firm, clean floor surface and the extension cord for the light and the clippers. Heather, my truly remarkable original mama sheep (who gave birth to more than 35 beautiful lambs before she finally died at the age of 15) was watching me, and I was telling her that tomorrow was shearing day. The next morning I went down to the barn shortly before Bob the shearer was to arrive. To my surprise, I found all 12 of the adult sheep stuffed like canned sardines, and as calm, inside the 4-by-8-foot lambing pen - but the gate was open. In previous years Bob had, with quite a bit of effort I might add, stuffed them all into this pen because it was then easy for him to pick out one sheep at a time for shearing, so I figured that Bob must have arrived early and already put them into the lambing pen. But then Bob arrived, and right away thanked me for putting them in already. Well, I hadn't done it, and Bob hadn't done it, and certainly no other person could have done it. Heather just stood there inside the opening of the bulging lambing pen, surrounded by the tight cluster of all her children, grandchildren, great-grandchildren, nieces and nephews, and her straight gaze leveled right into ours clearly told us: "I did it". I would have given anything to have been able to watch her accomplish that amazing feat.

Now, you may wonder after this story how sheep could possibly fit into food production. Let me try to explain. I believe in enabling animals to live in optimum freedom and happiness. This means letting the rams and the ewes live together to enjoy their natural relationships. This of course leads to lambs, which in turn brings joy to them all, and us. Lambs mean more and more sheep. If that continues beyond the carrying capacity of the land, the result is soil erosion and decreased nutrition from the pastures, therefore hunger and decreased happiness among the sheep, followed by decreased health.

One of the ways to prevent overpopulation is to keep the rams separated from the ewes. I tried that one year, and I can tell you it was no fun for anyone. There was constant frustration and stress. The rams tried to kill each other and were always working to knock down the fencing. We were always fixing the fencing, until finally at the height of the mating season nothing could hold the rams back from joining the ewes. The result was yet another generation of some 12 lambs five months later.

The other way to prevent overpopulation is the one I practice and believe is the best: allowing them the freedom to live together, which leads to their optimum happiness, and then controlling their numbers by killing some.

This is by far the most difficult task of the year. Selecting who goes is excruciatingly painful, but the rule is to keep those who are the healthiest and have the best quality wool, which in turn strengthens and improves the whole flock. Those who are selected to go are trucked to a small slaughter business where they are handled and killed humanely. A week later the boxes of neatly wrapped, labeled and frozen meat arrive, along with the salted skins. I put the meat in the freezer, and, after letting the skins dry for a couple of weeks, I send them to Bucks County Fur Products Quakerstown, a small tannery in Pennsylvania, where the messy raw pelts are miraculously transformed into the most heavenly soft fluffy skins, creamy white, tan, golden brown, gray-brown, and deep chocolate or whatever the color of the sheep who were chosen to be reincarnated.

Anyone who has ever sat or lain down on a sheep skin knows how wonderful it is. Sheep skins have a remarkable ability to keep the body floating on a cushion of air, and therefore they are widely used for premature babies and the long-term bedridden.

I love to cook, sometimes, and the meal that invariably generates the highest rating from family and friends is roast lamb. I cut off some of the fat and lay the leg in a roasting pan, put in an inch of water and let it brown on both sides in a 350 degree oven, with no lid, for about 60 minutes. Then I add more water so it is 2 inches deep, plus rosemary, thyme and marjoram, teriyaki sauce, pepper, garlic, onions and carrots. Then it is roasted, covered, at 300 degrees for 2 hours, after which I let it cool and put it in the refrigerator overnight. A couple of hours before dinner I remove all the fat which has hardened on top of the broth. The pan is returned to the oven, with more carrots and lots of little whole white onions, to cook for another hour at 300 degrees, covered, turning the meat every once in a while. Keep the liquid level a little over an inch. Finish off by removing the lid and roasting yet another 30 minutes. It is served with roasted or mashed potatoes. It does indeed make a memorable meal, so tender, so good and so good for you. It is just as delicious five days later and freezes very well provided no carrots or potatoes are included.

Whether relaxing on a sheep skin or feasting on roast lamb, we give thanks to the beings that they were, and with awe we contemplate the miracle and mystery of the continual cycles of life and death and the question that is probably as old as Homo sapiens: "What does happen to the spirit/mind/soul after death?"

Rafe Brown

Pastoral splendor.

Annika Edey

Rafe carrying lamb.

PART SEVEN
RECOMMENDED RESOURCES

Favorite Seed Varieties for Solviva Salad

Lettuces:

Red Saladbowl, Red Sails, Merveille de Quatre Saison, Buttercrunch, Rouge Grenobloise, Little Gem, Lollo Rossa, Royal Red, Green Wave, Ruby, French Antina, Vulcan, Juliet, Blackseeded Simpson, Saladbowl, Rosalita, Victoria, Romance, Ruben's Romaine, Verano, Waldman's, Grand Rapids, Prizehead, Sierra, Nancy.

Other Greens:

Sorrel de Belleville, Mache, Vates Blue Kale, Ruby Swiss Chard, Curly Endive, Full-heart Endive, Osaka Mustard, Radicchio Treviso, Arugula, Tah Tsai, Mizuna, Talltop Beet Greens.

Herbs:

Dill, Fennel, Cilantro, Sweet Basil, Chives, Parsley.

Edible Flowers:

Nasturtium, Alyssum, Pineapple Sage, Borage, Lime Geranium, Violets, Johnny-Jump-Up, Fennel, Cilantro, Chives, Garlic Chives, Calendula.

Seed Companies

Johnny's Seeds, Foss Hill Road, Albion, ME 04910 • (207) 437-4301
Stokes' Seeds, Box 548, Buffalo, NY 14240 • (800) 263-7233
Shepard's Garden Seeds, 30 Irene Street, Torrington, CT 06790 • (860) 482-3638
Seeds of Change, P.O. Box 15700, Santa Fe, NM 87506 • (505) 438-8080
The Cook's Garden, P.O.Box 65, Londonderry, VT 05148 * (802) 824-3400
Vermont Bean Seed Co, Garden Lane, Fair Haven, VT 05743

Beneficial Insects

Rincon-Vitova Insectaries, P.O.Box 1555, Ventura, CA 93002 • (805) 643-5407
Arbico, P.O.Box 4247 CRB, Tucson, AZ 85738-1247 • (800) 827-2847
Gardens Alive!, 5100 Schenley Place, Lawrenceburg, IN 47025 • (812) 537-8650

Sustainable Technologies

Sunelco, P.O.Box 787, Hamilton, Montana 59840 • (800) 338-6844
Real Goods, 555 Leslie Street, Ukiah, CA 95482 • (800) 762-7325
Woven Poly Plastic: Northern Greenhouse Sales, Box 42, Neche, N. Dakota 58265 • (204) 327-5540
Earthworms: Georgia Worm farms, P.O.Box 164, Dawson, GA 31742

Recommended Tool for Falling in Love with Life on Earth:

A Magnifying Lens, 3x to 10x. Closely observe at least 10 objects each day, a flower, a water drop, your skin, a bug... Perfect gift for each of your children, your friends, your enemies.

Good Source: Edmond Scientific Co., 101 E. Gloucester Pike, Barrington, NJ 08007 • (609) 547-8880

MOST RECOMMENDED READING

Earth in the Balance, by Al Gore. Houghton-Mifflin.
Soft Energy Paths, by Amory Lovins. Harper.
Energy Unbound, Amory Lovins et al. Sierra Club Books.
HOPE, Human and Wild, by Bill McKibben. Little Brown.
The Death of Common Sense, Philip Howard. Random House.
Diversity of Life, by E.O. Wilson. Belknap Harvard.
Silent Spring, by Rachel Carson
The Next One Hundred Years: Shaping the Fate of our Living Earth, by Jonathan Weiner. Bantam Books
Integral Urban House, by Olkowski, et al. Farrallones Institute. Sierra Club Books.
Restoring the Earth: Visionary Solutions from the Bioneers, Kenny Ausubel. Kramer Press

FURTHER RECOMMENDED READING

Billion and Billions: Thoughts on Life and Death at the Brink of the Millenium, Carl Sagan.
Pills, Pesticides and Profits, edited by Ruth Norris. North River Press
Toxic Deception: How the Chemical Industry Manipulates Science, Bends Laws, and Endangers Your Health, Fagin, Lavelle & Center for Public Integrity. Birch Lane Press.
Whole Earth Catalog, edited by Stewart Brand and J. Baldwin. Doubleday
Encyclopedia of Organic Gardening, Rodale Press
Mending the Earth: a World for our Grandchildren. Rothkrug & Olson. N. Atlantic Books.
Secret Life of Plants, Thompkins and Bird. Harper Row.
The Toilet Papers, by Sym Van der Ryn. Capra Press.
The Passive Solar Energy Book, by Edward Mazria. Rodale.
Solar Home Book, by Bruce Anderson. Cheshire Books.
Ecologue, by Bruce Anderson. Prentice Hall Press.
Bountiful Solar Greenhouse, by Shane Smith. Muir Publications.
Home-Made Money: How to Save Energy and Dollars in your Home, by Richard Heede, Rocky Mountain Institute
Photovoltaics: Sunlight to Electricity in One Step, Ron Maycock and Edward Stirewalt. Brick House Press
Energy Future, by Daniel Yergin and Robert Stobaugh. Random House.

PERIODICAL:

Growing Edge Magazine, (503) 757-0027
Amicus Journal, NRDC, (212) 727-2700
In Business, (610) 967-4135
Worldwatch Institute Journal, (800) 825-0061
Organic Gardening, (610) 967-5171
Sanctuary, Massachusetts Audubon Society, (617) 250-9500
Harper's Magazine, (212) 614-6500
Green Guide, (888) eco-info

Blueprints and Consulting are available

for the following

Solviva Solar-Dynamic Bio-Benign Designs:

- The Greenhouse, including the Animal Quarters, Raised Bed System, Hanging Growtubes, Waterwalls, Earthlung Filter, and Underground Firechamber for Pool Heating.
- The Walk-In Coldframe or Growshed.
- The Homes, Schools and Businesses, including the Solar Heating/Cooling Roof and Long-Term Deep-Ground Solar-Storage System
- The Backyard Food Factory
- The Backyard Home Office
- The Solar Powerplant Garage
- The Compostoilet
- The Solar Compost Pasteurization System
- The Graywater Purification and Irrigation System

For further information, please contact us at:

SOLVIVA COMPANY

RFD 1 Box 582, Vineyard Haven, MA 02568

Tel/Fax: (508) 693-3341

E-mail: solviva@vineyard.net

❦ ❦ ❦

An Invitation to Help Create the Next Solviva Book

You are invited to contribute information from around the world about problems or solutions regarding wastewater, solid wastes, heating, cooling, electricity, transportation or food production. It can be from your own experience, or it can be something you heard or read about. Please state the source of the information, because everything must be verifiable. Your contribution may be included in my next book, and will help in the struggle to improve the quality of all life on Earth.

❦ ❦ ❦

TABLE OF CONVERSION FACTORS

Because different conventions historically have been used to measure various quantities, the following tables have been compiled to sort out the different units. This first table identifies the units typically used for describing a particular quantity. For example, speed might be measured in "miles/hour".

Most quantities can also be described in terms of the following three basic dimensions:

length	L
mass	M
time	T

For example, speed is given in terms of length divided by time, which can be written as "L/T". This description, called "dimensional analysis", is useful in determining whether an equation is correct. The product of the dimensions on each side of the equal sign must match. For example:

$$\text{Distance} = \text{Speed} \times \text{Time}$$
$$L = L/T \times T$$

The dimension on the left side of the equal sign is length, L. On the right side of the equal sign, the product of L/T times T is L, which matches the left side of the equation.

The second table is a Conversion Table, showing how to convert from one set of units to another. It might be necessary to take the reciprocal of the conversion factor or to make more than one conversion to get the desired results.

Measured Quantities and Their Common Units

Length(L)	Area(L^2)	Volume(L^3)
mile(mi.)	sq. mile(mi^2)	gallon(gal.)
yard(yd.)	sq. yard(yd^2)	quart(qt.)
foot(ft.)	sq. foot(ft^2)	pint(pt.)
inch(in.)	sq. inch(in^2)	ounce(oz.)
fathom(fath.)	acre	cu. foot(ft^3)
kilometer(km.)	sq. kilometer(km^2)	cu. yard(yd^3)
meter(m.)	sq. meter(m^2)	cu. inch(in^3)
centimeter(cm.)	sq. centimeter(cm^2)	liter(l)
micron(μ)		cu. centimeter(cm^3)
angstrom(Å)		acre-foot
		cord(cd)
		cord-foot
		barrel(bbl.)

Mass(M)	Speed(L/T)	Flow Rate(L^3/T)
pound(lb.)	feet/minute (ft./min.)	cu. feet/min.
ton(short)	feet/sec.	cu. meter/min.
ton(long)	mile/hour	liters/sec.
ton(metric)	mile/min.	gallons/min.
gram(g.)	kilometer/hr.	gallons/sec.
kilogram(kg.)	kilometer/min.	
	kilometer/sec.	

Pressure($M/L/T^2$)	Energy(ML^2/T^2)	Power(ML^2/T^3)
atmosphere(atm.)	British thermal unit(Btu.)	Btu./min.
pounds/sq. inch(psi)	calories(cal.)	Btu./hour
inches of mercury	foot-pound	watt
cm. of mercury	joule	joule/sec.
feet of water	kilowatt-hour (kw-hr.)	cal./min.
	horsepower-hour (hp.-hr.)	horsepower(hp.)

Time(T)	Energy Density(M/T^2)	Power Density(M/T^3)
year(yr)	calories/sq. cm.	cal./sq. cm./min.
month	Btu./sq. foot	Btu./sq. foot/hr
day	langley	langley/min.
hour(hr.)	watthr./sq. foot	watt/sq. cm.
minute(min.)		
second(sec.)		

Table of Conversion Factors

MULTIPLY	BY	TO OBTAIN:
Acres	43560	Sq. feet
"	0.004047	Sq. kilometers
"	4047	Sq. meters
"	0.0015625	Sq. miles
"	4840	Sq. yards
Acre-feet	43560	Cu. feet
"	1233.5	Cu. meters
"	1613.3	Cu. yards
Angstroms(Å)	1×10^{-8}	Centimeters
"	3.937×10^{-9}	Inches
"	0.0001	Microns
Atmospheres(atm.)	76	Cm. of Hg(0°C)
"	1033.3	Cm. of H_2O(4°C)
"	33.8995	Ft. of H_2O(39.2°F)
"	29.92	In. of Hg(32°F)
"	14.696	Pounds/sq. inch(psi)
Barrels(petroleum, U.S.)(bbl.)	5.6146	Cu. feet
"	35	Gallons(Imperial)
"	42	Gallons(U.S.)
"	158.98	Liters
British Thermal Unit(Btu)	251.99	Calories, gm
"	777.649	Foot-pounds
"	0.00039275	Horsepower-hours
"	1054.35	Joules
"	0.000292875	Kilowatt-hours
"	1054.35	Watt-seconds
Btu/hr.	4.2	Calories/min.
"	777.65	Foot-pounds/hr.
"	0.0003927	Horsepower
"	0.000292875	Kilowatts
"	0.292875	Watts(or joule/sec.)
Btu/lb.	7.25×10^{-4}	Cal/gram
Btu/sq. ft.	0.271246	Calories/sq. cm. (or langleys)
"	0.292875	Watt-hour/sq. foot
Btu/sq. ft./hour	3.15×10^{-7}	Kilowatts/sq. meter
"	4.51×10^{-3}	Cal./sq. cm./min(or langleys/min)
"	3.15×10^{-8}	Watts/sq. cm.
Calories(cal.)	0.003968	Btu.
"	3.08596	Foot-pounds

"	1.55857×10^{-6}	Horsepower-hours
"	4.184	Joules (or watt-sec s)
"	1.1622×10^{-6}	Kilowatt-hours
Calories, food unit (Cal.)	1000	Calories
Calories/min.	0.003968	Btu/min.
"	0.06973	Watts
Calories/sq. cm.	3.68669	Btu/sq. ft.
"	1.0797	Watt-hr/sq. foot
Cal./sq. cm./min.	796320.	Btu/sq. foot/hr.
"	251.04	Watts/sq. cm.
Candle power (spherical)	12.566	Lumens
Centimeters(cm.)	0.032808	Feet
"	0.3937	Inches
"	0.01	Meters
"	10.000	Microns
Cm. of Hg(0°C)	0.0131579	Atmospheres
"	0.44605	Ft. of H_2O(4°C)
"	0.19337	Pounds/sq. inch
Cm. of H_2O(4°C)	0.0009678	Atmospheres
"	0.01422	Pounds/sq. inch
Cm./sec.	0.032808	Feet/sec.
"	0.022369	Miles/hr.
Cords	8	Cord-feet
"	128(or 4×4×8)	Cu. feet
Cu. centimeters	3.5314667	Cu. feet
"	0.06102	Cu. inches
"	1×10^{-6}	Cu. meters
"	0.001	Liters
"	0.0338	Ounces(U.S. fluid)
Cu. feet(ft.^{3})	0.02831685	Cu. meters
"	7.4805	Gallons(U.S., liq.)
"	28.31685	Liters
"	29.922	Quarts(U.S., liq.)
Cu. ft. of H_2O (60°F)	62.366	Pounds of H_2O
Cu. feet/min.	471.947	Cu. cm./sec.
Cu. inches(in.^{3})	16.387	Cu. cm.
"	0.0005787	Cu. feet
"	0.004329	Gallons(U.S., liq.)
"	0.5541	Ounces(U.S., fluid)
Cu. meters	1×10^6	Cu. centimeters
"	35.314667	Cu. feet
"	264.172	Gallons(U.S., liq.)
"	1000	Liters
Cu. yard	27	Cu. feet
"	0.76455	Cu. meters
"	201.97	Gallons(U.S., liq.)
Cubits	18	Inches
Fathoms	6	Feet
"	1.8288	Meters
Feet(ft.)	30.48	Centimeters
"	12	Inches
"	0.00018939	Miles(statute)
Feet of H_2O(4°C)	0.029499	Atmospheres
"	2.2419	Cm. of Hg(0°C)
"	0.433515	Pounds/sq. inch
Feet/min.	0.508	Centimeters/second
"	0.018288	Kilometers/hr.
"	0.0113636	Miles/hr.
Foot-candles	1	Lumens/sq. foot
Foot-pounds	0.001285	Btu.
"	0.324048	Calories
"	5.0505×10^{-7}	Horsepower-hours
"	3.76616×10^{-7}	Kilowatt-hours

Furlong	220	Yards
Gallons(U.S., dry)	1.163647	Gallons(U.S., liq.)
Gallons(U.S., liq.)	3785.4	Cu. centimeters
"	0.13368	Cu. feet
"	231	Cu. inches
"	0.0037854	Cu. meters
"	3.7854	Liters
"	8	Pints(U.S., liq.)
"	4	Quarts(U.S., liq.)
Gallons/min.	2.228×10^{-3}	Cu. feet/sec.
"	0.06308	Liters/sec.
Grams	0.035274	Ounces(avdp.)
"	0.002205	Pounds(avdp.)
Grams-cm.	9.3011×10^{-8}	Btu.
Grams/meter^2	3.98	Short ton/acre
"	8.92	lbs./acre
Horsepower	42.4356	Btu./min.
"	550	Foot-pounds/sec.
"	745.7	Watts
Horsepower-hrs.	2546.14	Btu.
"	641616	Calories
"	1.98×10^6	Foot-pounds
"	0.7457	Kilowatt-hours
Inches	2.54	Centimeters
"	0.83333	Feet
In. of Hg(32°F)	0.03342	Atmospheres
"	1.133	Feet of H_2O
"	0.4912	Pounds/sq. inch
In. of Water(4°C)	0.002458	Atmospheres
"	0.07355	In. of Mercury(32°F)
"	0.03613	Pounds/sq. inch
Joules	0.0009485	Btu.
"	0.73756	Foot-pounds
"	0.0002778	Watt-hours
"	1	Watt-sec.
Kilo calories/gram	1378.54	Btu/lb
Kilograms	2.2046	Pounds(avdp.)
Kilograms/hectare	.893	lbs/acre
Kilograms/hectare	.0004465	Short ton/acre
Kilometers	1000	Meters
"	0.62137	Miles(statute)
Kilometer/hr.	54.68	Feet/min.
Kilowatts	3414.43	Btu./hr.
"	737.56	Foot-pounds/sec.
"	1.34102	Horsepower
Kilowatt-hours	3414.43	Btu.
"	1.34102	Horsepower-hours
Knots	51.44	Centimeter/sec.
"	1	Mile(nautical)/hr.
"	1.15078	Miles(Statute)/hr.
Langleys	1	Calories/sq. cm.
Liters	1000	Cu. centimeters
"	0.0353	Cu. feet
"	0.2642	Gallons(U.S., liq.)
"	1.0567	Quarts(U.S., liq.)
Lbs./acre	.0005	Short ton/acre
Liters/min.	0.0353	Cu. feet/min.
"	0.2642	Gallons(U.S., liq.)/min.
Lumens	0.079577	Candle power(spherical)
Lumens(at 5550Å)	0.0014706	Watts
Meters	3.2808	Feet
"	39.37	Inches
"	1.0936	Yards
Meters/sec.	2.24	Miles/hr.
Micron	10000	Angstroms
"	0.0001	Centimeters
Miles(statute)	5280	Feet

"	1.6093	Kilometers
"	1760	Yards
Miles/hour	44.704	Centimeter/sec.
"	88	Feet/min.
"	1.6093	Kilometer/hr.
"	0.447	Meters/second
Milliliter	1	Cu. centimeter
Millimeter	0.1	Centimeter
Ounces(avdp.)	0.0625	Pounds(avdp.)
Ounces(U.S., liq.)	29.57	Cu. centimeters
"	1.8047	Cu. inches
"	0.0625(or 1/16)	Pint(U.S., liq.)
Pints(U.S., liq.)	473.18	Cu. centimeters
"	28.875	Cu. inches
"	0.5	Quarts(U.S., liq.)
Pounds(avdp.)	0.45359	Kilograms
"	16	Ounces(avdp.)
Pounds of water	0.01602	Cu. feet of water
"	0.1198	Gallons(U.S., liq.)
Pounds/acre	0.0005	Short ton/acre
Pounds/sq. inch	0.06805	Atmospheres
"	5.1715	Cm. of mercury(0°C)
"	27.6807	In. of water(39.2°F)
Quarts(U.S., liq.)	0.25	Gallons(U.S., liq.)
"	0.9463	Liters
"	32	Ounces(U.S., liq.)
"	2	Pints(U.S., liq.)
Radians	57.30	Degrees
Sq. centimeters	0.0010764	Sq. feet
"	0.1550	Sq. inches
Sq. feet	2.2957×10^{-5}	Acres
"	0.09290	Sq. meters
Sq. inches	6.4516	Sq. centimeters
"	0.006944	Sq. feet
Sq. kilometers	247.1	Acres
"	1.0764×10^7	Sq. feet
"	0.3861	Sq. miles
Sq. meters	10.7639	Sq. feet
"	1.196	Sq. yards
Sq. miles	640	Acres
"	2.788×10^7	Sq. feet
"	2.590	Sq. kilometers
Sq. yards	9(or 3×3)	Sq. feet
"	0.83613	Sq. meters
Tons, long	1016	Kilograms
"	2240	Pounds(avdp.)
Tons(metric)	1000	Kilograms
"	2204.6	Pounds(avdp.)
Tons, metric/hectare	0.446	Short ton/acre
Tons(short)	907.2	Kilograms
"	2000	Pounds(avdp.)
Watts	3.4144	Btu./hr.
"	0.05691	Btu./min.
"	14.34	Calories/min.
"	0.001341	Horsepower
"	1	Joule/sec.
Watts/sq. cm.	3172	Btu./sq. foot/hr.
Watt-hours	3.4144	Btu.
"	860.4	Calories
"	0.001341	Horsepower-hours
Yards	3	Feet
"	0.9144	Meters

And finally,
to remind us,
in among the seriousness of it all,
how funny it all is, we all are,
and as a grand finale to this book,
I offer this conversion chart,
gleaned from that extraordinary
global bulletin, the Internet.
(God bless us all
as we learn to surf
into the next Millenium...)

The Ultimate Metric Conversion Chart:

10^{12} Microphones = 1 Megaphone

10^6 bicycles = 2 megacycles

500 millinaries = 1 seminary

2000 mockingbirds = two kilomockingbirds

10 cards = 1 decacards

1/2 lavatory = 1 demijohn

10^{-6} fish = 1 microfiche

453.6 graham crackers = 1 pound cake

10^{12} pins = 1 terrapin

10^{21} picolos = 1 gigolo

10 rations = 1 decoration

100 rations = 1 C-ration

10 millipedes = 1 centipede

3 1/3 tridents = 1 decadent

5 holocausts = 1 Pentacost

10 monologs = 5 dialogues

5 dialogues = 1 decalogue

2 monograms = 1 diagram

8 nickels = 2 paradigms

2 snake eyes = 1 paradise

2 wharves = 1 paradox

NOTES

NOTES

NOTES

NOTES

NOTES

NOTES

NOTES

NOTES